SCIENCE AND INDUSTRY
IN THE NINETEENTH CENTURY

SCIENCE AND INDUSTRY
IN THE
NINETEENTH CENTURY

by

J. D. BERNAL, F.R.S.

INDIANA UNIVERSITY PRESS
BLOOMINGTON AND LONDON

Published in Canada by Fitzhenry & Whiteside Limited,
Don Mills, Ontario

Library of Congress catalog card number: 71-89516
SBN: 253–20128–4

Manufactured in the United States of America

PREFACE

THIS book contains two essays of very different length and scope but with closely related subjects. A longer one covering in a general way the relations of science and technology in the nineteenth century and a shorter one analysing in detail just one such interaction—the discovery of molecular asymmetry by Pasteur in 1848, his first and in some ways his greatest scientific discovery.

'Science and Industry' is the result of some studies I undertook two or three years ago. At the time I was preparing a contribution to a collection of essays dealing with various aspects of the relations between science and society to be published on behalf of the International Union for the History of Science.[1]

I found myself obliged, in order to begin to understand the links between industry and science, to examine a few specific cases in detail. Once I had started doing this I became so fascinated with the stories themselves and the unexpected connections they revealed that it became difficult to know where to stop and how to avoid becoming confused by the wealth of material that came to light. There was clearly far more than I could put into an article and

[1] Published in *Centaurus*, Vol. 3, Nos. 1-2, 1953.

yet a real study of the subject would need several volumes and take a lifetime, even if I were the person to do it.[1]

I decided therefore to cut the matter short and to publish the brief and schematic account I had prepared in the hope that it would stir others to further studies in this field. I do not expect either my facts or my conclusions to pass unchallenged, but criticism and controversy cannot fail to focus more light on an aspect of science and of history that demands in these days to be profoundly studied. For it is only by understanding the workings of science in the past that we can control it in the present and in the future.

'Molecular Asymmetry' was prepared as a lecture given in 1946 at the Congress of Commemoration of the Fiftieth Anniversary of the Death of Pasteur. I have included it because it deals, in greater detail than the other examples I have given, with the actual history of a great discovery, its antecedents and its consequences. It also serves to fill a gap in the account of the general contribution of Liebig and Pasteur which I have given in Chapter III of my first essay.

My approach in the two cases is naturally a different one. In the first essay I was concerned with the broad interactions of scientific and technological factors, with somewhat of an emphasis on the latter. In the second the centre of interest is rather in the scientific chain of causa-

[1] We may hope to see the technical part of the story in the projected *History of Technology* to be edited by Dr. Charles Singer, and of course there exist many good histories of science. But a history of the relations of science and technology can no more be made by simple juxtaposition than can a treatise on Chinese metaphysics.

tion, and the economic factors, though they exist, are less in evidence. Indeed, this first of Pasteur's discoveries, undertaken in the École Normale, was naturally far removed from practical considerations. The response to financial, patriotic, and humanitarian needs that were to guide Pasteur through his later discoveries on fermentation and on the diseases of animals and men were not yet in evidence.

This crucial discovery is one of the greatest of the nineteenth century. It is a classical case of the convergence of ideas from various fields—chemical, crystallographic, and physical—to reveal a new property of matter, molecular asymmetry. All the relevant facts, save the one specific observation which brought it to light, had been established years before, and that one fact—the two kinds of hemihedry of Seignette salt—could have been observed by dozens of competent crystallographers of the day. The problem which the young Pasteur deliberately set himself was not an obscure and neglected one; it had occupied the best brains of European science for the best part of twenty years. Yet they failed to solve it and Pasteur in his very first research did so easily. Why? Because he was already a master of the techniques of chemistry and crystallography? Because he was remarkably clear-headed, could formulate the problem and was free from preconceptions and preoccupations? Was there any other reason? I had studied the history of the discovery in the works of Pasteur's predecessors and in his own writings as best I could when I was asked to lecture on it at the Paris conference. I thought I had mastered the chief lines of evidence and the thoughts that led him to his conclusion. I arrived in Paris with my lecture ready. It was only then that I had the good fortune, rare in a lifetime, to come across Pasteur's original and unedited notebooks of the discovery where another story, no less an achievement but far more illuminating, was revealed. Pasteur had other clues which time had effaced. I recast my lecture rapidly—the matter required far more and deeper study than I could give it then, or have given it since. I could have rewritten my essay in the cooler light of scholarship, but I have preferred to leave it as it was because it presents in a way I could not again recapture my immediate reactions to the authentic record of a great discovery, buried as it had been for nearly a hundred years.[1]

A discovery is a beginning as much as an end in science, and from that of molecular asymmetry whole new sciences such as stereochemistry and bacteriology were to arise. In this essay I try to sketch in outline these and other consequences and to discuss why some of them were so long delayed. They represent a parallel in the history of ideas to the economic obstructions that held up the development of technical invention in the same period. The whole theme is one of a close-knit bunch of strands in the history of science and technology in the nineteenth century. It ramifies and links, before as well

[1] The words are not the same. The original lecture given in French, 'Dissymétrie Moléculaire', is to be found in the report of the *Congres des Sciences Pastoriennes pour le cinquantenaire de la mort de Louis Pasteur*, Union Nationale des Intellectuels, Paris, 1948. I must thank here Mr. Francis Aprahamian for making the translation and for much other help in preparing this book.

as after, with many other themes, but one knot holds it all together and makes it an entity in its own right—the central discovery of molecular asymmetry. The discovery cannot be separated from the man, a great figure, one of the greatest of the nineteenth century. In many ways he may stand as the type of the period, coming into scientific production in the year of crisis of 1848 and passing from the scene in 1896, effectively the last year of the nineteenth century in science, for it was the year of the discovery of X-rays and electrons that looks forward to the great new advance of the twentieth century. The two essays therefore, though composed on different occasions and for different reasons, turn out to be complementary. In one we can glimpse at and survey roughly the broad field of industrial and scientific relations, lighting it up here and there to emphasize some particularly revealing interaction. In the other the light is focused on one spot—the single discovery to which different clues in industry as well as in science lead and from which comes as wide a variety of consequences. Both approaches illustrate different aspects of one process and may serve to show the richness of the field which calls for survey and analysis and the relevance of such studies for the full understanding of the science and industry of our own times.

SCIENCE AND INDUSTRY
IN THE NINETEENTH CENTURY

Preface to the Second Edition

THE MAIN OBJECT of the first essay in this book, written in 1953, was to bring out the reciprocal relationships of science and technology in the nineteenth century and their links with the economic, political and cultural forces of that era. The passage of time has not lessened the importance of the study of these relationships: in fact it is now a key aspect of what has come to be known as the science of science.[1]

In recent years, government and industry, both socialist and capitalist, have set up special institutes or study groups to investigate such problems as the return on investment in research, the reasons for lags in the application of scientific discoveries, optimum investment in basic and applied research, etc. A re-issue of this book is, therefore, timely, and its excursion into the past may indicate a useful approach to current problems.

In this short summary it is not possible even remotely to do justice to the exciting developments that have taken place in the extraordinarily fruitful period of the last seventeen years or so since this book was published.

Perhaps one of the most outstanding instances of the ever-increasing reaction between society, science and technology has been the 'take-over' of nineteenth century 'Little Science' by 'Big Science.'[2] The greatest social and economic consequences have probably arisen from the production of nuclear power and weapons, rockets, guidance and communications systems, which include space travel, and, the pivot of them all, the computer, surely the greatest mechanical, mathematical and electronic invention so far of this or any other century.

The introduction into science and technology of the quantum theory has transformed the scene of energy levels and made possible such concepts as nuclear energy, fission and fusion, which are now a reality and producing their own chain reaction in industry and in war. At the same time the nineteenth century age of steel has been transformed by the use of powder methods and of low-temperature and powder reactions in place of the furnace.

The importance of Pasteur's discovery of Molecular Asymmetry, dealt with in the second essay, has also grown with the years and it now dominates the whole of chemistry. Since it was written, we have witnessed one of the greatest triumphs of a similar convergence of ideas from the various scientific—and, even more, technological—fields, in the elucidation of the structure of DNA and the founding of the new scientific disciplines of molecular biology and stereochemical reactions.

The introduction of the electron microscope, both transmission and scanning, has made possible revelations of the nature and function of viruses and other minute organisms as well as the nature of the cell and its organelles. And we are only at the beginning of the possibilities for the adaptation and use of the electron microscope. In the next few years it will undoubtedly play a vital part in unravelling what is probably the most exciting problem of modern science—that of the origin of life itself.   J. D. BERNAL

January 1969

NOTES

1. *The Science of Science,* ed. M. Goldsmith and A. L. Mackay, Souvenir Press, Ltd., London, W.C.1., and Ryerson Press, Toronto 2, Canada, 1964.

2. *Little Science, Big Science,* Derek de Solla Price, Columbia University Press, 1963.

CONTENTS

Antecedents and consequences of Pasteur's discovery of molecular asymmetry

SCIENCE AND INDUSTRY
IN THE NINETEENTH CENTURY

Chapter I

INTRODUCTION

WE who live in an age where science is recognized as a means of life or death, cannot fail to see all around us the consequences and even the instruments of science. That very fact, however, makes it extremely difficult to disentangle science from the social and economic factors with which it is entwined. Scientists themselves are at a loss to know how far their responsibility extends into the consequences for good or evil of discoveries and applications often made more collectively than individually. There is no recognized means of assessing the amount of a community's resources that should go to science, how it should be apportioned or indeed whether the whole matter should not be left to chance, as it has been so largely in the past.

These are not academic questions—we need the solutions to deal with our day-to-day problems and for planning the most immediate future. Yet they can be solved only by a study which takes account of how the present grew out of the past, for science and technology are preeminently traditional social institutions, depending for their very existence on an accumulated stock of facts and methods to a far larger degree and far more consciously than do the arts. That is why it may be of some value to examine the relations of science and technology, or, more widely, of science and industry, in an era like the nineteenth century when those relations were simpler than they are now, but yet one not so distant that we cannot appreciate from our own experiences the significance of its main movements.

Short as has been the gap in years we are now getting far enough away from the nineteenth century to be able to see its achievements in science and technology in the wide perspective of history. Nevertheless, the task of finding the relations between them is by no means an easy one. Science is still a somewhat unfamiliar part of social life and those outside its disciplines find it hard to realize the changes that have taken place in them. Consequently, many intelligent non-scientific people still think of science as it appeared to be in the nineteenth century, as the product of individual efforts of men of genius, instead of, as it now is, a highly organized new profession closely linked with industry and government. On the other hand many scientists of today, outside the older centres of learning where the ways of the past still linger, find it difficult to grasp the uncoordinated and amateur character of nineteenth-century science with little formal teaching and without research laboratories or research funds. It is almost as difficult in an age of vast engineering and chemical factories, each furnished with its own research department, to recall the intimate traditional and practical character of the old workshops and forges from which the modern giants are descended.

Introduction

In fact the nineteenth century was as different from the twentieth as it was from the eighteenth. It was above all a period of expansion—expansion of population, of manufacture, of trade and of knowledge. In their time these increases seemed unlimited, they were taken to herald the achievement of a universal Progress that was reflected in the world of nature itself by the great generalization of Evolution. All this also seemed to be a most natural, as well as desirable, state of affairs. With the advent of free-trade capitalism in the mid-century, economics was deemed to have found its true laws which the ignorance and superstition of earlier ages had hidden from sight. By abandoning all restrictions a *laisser-faire* Liberalism would achieve the best distribution of wealth by the automatic operation of the laws of the market.

What actually happened, as we know, was very different. Far from producing peaceful progress, the nineteenth century ushered in the transitional period of upheaval and violence of the twentieth century. We can see it now as a period of material and social *preparation* for a far more radical revolution of production, distribution and government. This revolution was an implicit consequence of the great new productive forces released by the scientific and technical advances achieved in principle in the eighteenth century but first realized on a large scale in practice only in the nineteenth.

Then, for the first time, it becomes possible to deal with the relations of science and technology in such a relatively short space as a hundred years. The pace of application of scientific discovery was speeding up sufficiently for the effect of discoveries made early in the century to be appreciable by its end. The use of the electric current, discovered just before the beginning of the century, was appreciable in the telegraph in the fifties though it was only beginning to be used as electric light and power in the nineties. In general the industry of the nineteenth century depended on the twin scientific and technical achievements of the late eighteenth century: the development of the steam engine and the establishment of a rational, quantitative chemistry. The greatest achievements of the physical sciences of the nineteenth century—the doctrine of the conservation of energy and the interchangeability of its various forms; the sciences of thermodynamics and electrodynamics—drew their inspiration from the study of practical sources of power and arose from the needs of transport and communication. Their full utilization as the basis of a rational chemical and electrical industry had to wait till the twentieth century.

For the historian a century is necessarily a most arbitrary and often inconvenient division of time; it is doubly so when the histories of two different human activities have to be considered together. A longer or shorter period, say from 1760 to 1914 or 1820 to 1870, would have advantages in considering the history of technology, the former bridging the whole of the Industrial Revolution before the period of mass production, the latter concentrating on the characteristic nineteenth-century achievements of the railway, steamship and telegraph. In the history of science, the limits are more difficult to define at the beginning and easier at the ened. The era might start with Hales and Black and the beginning of the pneumatic revolution in the mid-eighteenth century or, alternatively, considering that revolution as already complete, with 1831, the year of the foundation of the British Association and the discovery of electromagnetic induction by Faraday. The end of the era is definite enough, at least for physical science, because it is marked by a break-through to a new and unsuspected realm of experience. This was the discovery of X-rays by Röntgen in 1895, followed almost immediately by that of the electron and radioactivity and leading to the theory of atomic structure, the central feature of twentieth-century science. The choice taken here is to limit the century in the beginning by a social fact, the outbreak of the French Revolution in 1789, and to end it by 1895, a date fixed for scientific reasons which roughly coincides with a turning-point in the development of capitalism when the division of the world into empires was completed and preparation for a new era of wars was consciously beginning. These limitations of time will not preclude a certain casting backward for origins or looking forward for consequences.

The aim of this essay is to bring out by the study of actual examples the close and necessary connections between technical developments and the advance of scientific knowledge. These connections are not limited to any period of history but have a critical importance in the nineteenth century.[1] Before the Industrial Revolution,

[1] This was evident enough to far-sighted men of the time, especially to those who could see below the surface of the chaos of apparently unrelated discoveries and inventions. In 1844, almost at the outset of their career, Marx and Engels, attacking Feuerbach's idealist picture of mental progress, were writing: ' . . . the celebrated "unity of man with nature" has always existed in industry and has existed in varying forms in every epoch according to the lesser or greater development

7

science had been an affair of courts, gentlemen and scholars, and except for the arts of navigation and war it hardly affected ordinary life. The idea that it could do so was a vision enthusiastically acclaimed but as often derided.[1] By the twentieth century, on the other hand, the

of industry, just like the "struggle" of man with nature, right up to the development of his productive powers in a corresponding basis. Industry and commerce, production and the exchange of the necessities of life, themselves determine distribution, the structure of the different social classes and are, in turn, determined by these as to the mode in which they are carried on; and so it happens that in Manchester, for instance, Feuerbach sees only factories and machines where a hundred years ago only spinning-wheels and weaving-looms were to be seen, or in the Campagna of Rome he finds only pasture lands and swamps, where in the time of Augustus he would have found nothing but the vineyards and villas of Roman capitalists. Feuerbach speaks in particular of the perception of natural science; he mentions secrets which are disclosed only to the eye of the physicist and chemist: but where would natural science be without industry and commerce? Even this "pure" natural science is provided with an aim, as with its material, only through trade and industry, through the sensuous activity of men. So much is this activity, this unceasing sensuous labour and creation, this production, the basis of the whole sensuous world as it now exists, that, were it interrupted only for a year, Feuerbach would not only find an enormous change in the natural world, but would very soon find that the whole world of men and his own perceptive faculty, nay his own existence, were missing.'

K. Marx and F. Engels, The German Ideology,
London, 1938, p. 36.

[1] The advocacy of science as a means to 'the benefit and relief of the state and society of man', so eloquently preached in Francis Bacon's *Novum Organum* and *New Atlantis* and later in Sprat's *History of the Royal Society*, 1667, and Glanvil's *Plus Ultra*, 1668, were answered in Stubbes' *The Plus Ultra reduced to a Non Plus*, 1670, and, best known

interrelations of science and technique were consciously recognized. Whether constructive or destructive ends

of all, by Swift in *Gulliver's Travels*, 1726. In fact the hopes of the first projectors were not to be fulfilled in practice for over a century. Some reasons for this have been brought out recently in the detailed study by Hall, *Ballistics in the Seventeenth Century*, where he shows that the failure of the mathematicians and virtuosi of that time to transform technique was not at all through want of will, application, or understanding, but simply because the manual trades of the time did not admit of accuracy. If this was so in the vital field of gunnery, it must be admitted that the general state of manufacture until the nineteenth century could not in fact make use of the refined calculations which the mathematicians of the time could already make. A quotation from his opening chapter introduces very well the thesis of this essay:

'To say that the economic life of society in general, and processes of manufacture in particular, were unaffected by science until the beginning of the last century is scarcely an exaggeration. The seventeenth-century revolution in thought and method had moulded a science which was potentially capable of effecting profound changes in the means of production, and in fact many writers on science at the time found an important justification for the study of science in the fuller exploitation of natural resources, with the consequent enrichment of human life and alleviation of daily toil which it promised. But this promise was only fulfilled through the industrial and agrarian revolutions of the nineteenth century and the changes in the organization of economic activity which they brought about. In particular the sudden rise of engineering needs above the level of the carpenter and the blacksmith, the sudden realization that engineering skill in all its branches was fundamental to improvements in manufacture, transportation, agriculture and the means of making war, created a situation in which scientific knowledge and method not only could be, but must be, applied, while large-scale manufacture provided the means and incentive for the application of science.'

A. R. Hall, *Ballistics in the Seventeenth Century*, Cambridge, 1952, p. 1.

were in view there could be no doubt that the means employed for the advancement of techniques must be scientific. The transition between the dream and the reality was effected in the nineteenth century and it is therefore specially important to inquire how and why it occurred.

Before attempting to analyse in some detail the connections between the scientific and technical developments of the nineteenth century it is useful to give a general summary of the main trends which revealed themselves in science and technology separately.

Each field, the technical as much as the scientific, has its own inner coherence, not only in the logical unfolding of new discoveries on the basis of older researches and in the making of new inventions drawing on older technical advances, but also in their being in the hands of two largely distinct sets of men, the scientists and the engineers. At the beginning of the century the personal interaction was greatest, the engineers and scientists were the same men or were close friends, but the state of the sciences themselves provided only certain limited bridges between theory and practice. On the other hand towards its end the scientists and engineers, incorporated in their distinct societies and institutions, had drawn further apart, but by then the advance of science had made its intervention into techniques possible and indeed necessary over a large part of the field, while conversely the problems, the equipment and, not least, the funds of science were provided by industry.

Introduction

The main lines of scientific advance .

In science, the nineteenth century was the great period of specialization, as witness the formation of the separate scientific societies to supplement the older general academies such as the Royal Society. Each discipline followed its own line of development, they were not yet ready for the general unification of the sciences which is the major task of the twentieth century. Such unification as occurred lay inside each science; in physics with the great generalization of the electromagnetic theory of light; in chemistry with the union of organic and inorganic through the theory of valency. Both were achieved only towards the end of the century and both seemed to indicate a finality that was soon to prove illusory.

New trends in physics

By the beginning of the nineteenth century the specific field opened by the great Galilean-Newtonian union of mechanics and astronomy had been effectively worked out, though it still dominated academic science through its immense prestige. There was no longer either practical use or new scientific knowledge to be gained by following out the theory of gravitation to its ultimate conclusions. Leverrier and Adams' discovery of Neptune by this means was the greatest triumph, but at the same time the last effort, of classical astronomy.[1] When gravitation theory reappeared in the twentieth century with Einstein, it was in a very different physical context. But though New-

[1] In 1862 W. Thomson, on the advice of Tait, omitted all reference to Neptune in his introductory lecture to students on the ground that it had been 'ridden to death'. See S. P. Thompson's *Life of Lord Kelvin*, p. 244.

tonian mechanics might have little more to add in its own field, it was to give birth to Newtonian physics. Where the classical dynamics found its use was in the evolution of a mathematical language, in the hands of Lagrange, Fourier, Hamilton and Gauss, to describe the physical phenomena of a more generalized character, such as those of electricity and magnetism or, on a molecular scale, of the kinetic theory of gases and the foundations of thermodynamics.

Electricity

In physics the new era effectively began with the discovery by Galvani and Volta of the electric current. This was originally derived from the study of nerve physiology but for most of the century it was to be developed on a physical basis. The study of electricity provided an inexhaustible fund of new and exciting phenomena providing a lasting stimulus towards further experimentation of a qualitative kind, calling for an explanation by mathematical theory in quantitative terms, and, as we shall see in more detail later, opening up a sequence of inviting opportunities to commercial exploitation. All three threads are to be found in the culminating generalization of the century—Maxwell's electromagnetic theory of light—for this was based on the experiments of Faraday, on the theories of Fresnel and Gauss and on the relation of systems of electrical and magnetic units of Weber. It was in turn to lead through Hertz to the practical control of radiation of the twentieth century. The development and mastery of the electric current and of its magnetic and chemical manifestations was one great task of nineteenth-century physics.

Introduction

Thermodynamics

The other main branch of physics grew more directly out of the operation of the great eighteenth-century achievement—the steam engine—or the philosophical engine as it was so rightly called. The economic production and use of mechanical power was the inspiration of Carnot, Joule, Rankine and Thomson. It led, characteristically, first to the concept of maximum utilizable energy—the second law of thermodynamics—and then to the conception of the indestructibility and interchangeability of all forms of energy—the first law of thermodynamics. This was to bring its return to industry in the development of the internal combustion engine and the practice of refrigeration. Extended to chemistry by Clausius and Gibbs it was to be the basis of the rational chemical industry of the twentieth century. The electrical and thermal streams of nineteenth-century science, inspired by the new phenomena of atomic physics, were to come together in the generalizations of Planck and Einstein in the quantum theory of the twentieth century.

Chemistry

In chemistry it would be more logical to begin the story right back in the eighteenth century with the pneumatic revolution—the study of gases—which ushered in the oxygen theory of combustion through the logical analysis by Lavoisier of the experiments of Priestley, Scheele and Cavendish. But this story really belongs to the concentrated and combined intellectual and technical effort of the early Industrial Revolution. We can start the new century with the assumption that the period of purely traditional chemistry is over and that rational methods, based on a clear conception of chemical elements and the law of constancy of mass, can now be used. The additional keys of electrochemical decomposition and of the atomic theory were provided at the very opening of the century by Davy and Dalton. Nevertheless, so great was the variety of mineral, and even more of organic, substances that it took the best part of a hundred years to use these keys effectively to understand the structure of chemical compounds and reactions between them. And mere complexity was not the only obstacle. Another was the difficulty of shaking off the influence of the half-magical, half-metaphysical, chemical theories of the past, exemplified by the opposition to the atomic theory which was still maintained right to the end of the century.

At all stages, as we shall see, these advances are connected with the solutions of problems presented by a growing chemical industry and, in turn, give rise to new branches of that industry, such as those of coal-tar dyes or alkali manufacture. These were, however, but a small foretaste of the possible uses of chemistry. Throughout the nineteenth century, chemistry was essentially feeling its way, concentrating on the purification of natural substances and on their analysis. It ventured on the synthesis of only very simple molecules and attempted only small modifications of accessible substances. Despite the pioneer work of Berthelot, industrial chemical synthesis of complex molecules, starting from the elements themselves, was not to come till well into the twentieth century.

Closely linked with physics as it was in the beginning of the century, chemistry grew so fast that it became virtually autonomous and developed its own laws as the century progressed. And this autonomy persisted even though important instruments such as the polariscope and the spectroscope were borrowed from physics, as were also major guiding principles, such as those of thermo-dynamics and the kinetic theory of gases and solutions. No full union of chemistry and physics was, however, possible without some physical picture of the structure of the atom, which could come only in the twentieth century.

Biology

Although this essay is intended to deal primarily with physical science, it is impossible to achieve a balanced picture of scientific advance without reference to the enormous biological advances of the nineteenth century. These advances—the cell theory, the theory of evolution, the germ theory of fermentation and of disease, the eluci-dation of the main lines of the physiology of animals and plants—all contributed both to the general scientific atmosphere and to the direction of research in the physical sciences.[1] The devotion of Liebig to agricultural chemistry

[1] The unity of nineteenth-century physical and biological science was well expressed by Engels in 1885 in his review on Ludwig Feuerbach:

'But, above all, there are three great discoveries which have en-abled our knowledge of the interconnection of natural processes to advance by leaps and bounds: first, the discovery of the cell as the unit from whose multiplication and differentiation the whole plant and animal body develops—so that not only is the development and growth of all higher organism recognized to proceed according to a single general law, but also, in the capacity of the cell to change, the way is pointed out by which organisms can change their species and thus go through a more than individual development. Second, the

and to nutrition and that of Helmholtz to the physiology of the senses are examples of the linkage already occurring between the physical and biological sciences in the mid-nineteenth century and are a small-scale forecast of the greater integration of the present time.

Geology

With the biological sciences should be ranked, throughout the nineteenth century, the science of geology. This science, taking its rise appropriately with the theories of

transformation of energy, which has demonstrated that all the so-called forces operative in the first instance in inorganic nature—mechanical force and its complement, so-called potential energy, heat, radiation (light or radiant heat), electricity, magnetism and chemical energy—are different forms of manifestations of universal motion, which pass into one another in definite proportions so that in place of a certain quantity of the one which disappears, a certain quantity of another makes its appearance and thus the whole motion of nature is reduced to this incessant process of transformation from one form into another. Finally, the proof which Darwin first developed in connected form that the stock of organic products of nature surrounding us today, including mankind, is the result of a long process of evolution from a few original unicellular germs, and that these again have arisen from protoplasm or albumen which came into existence by chemical means.

'Thanks to these three great discoveries and the other immense advances in natural science, we have now arrived at the point where we can demonstrate as a whole the interconnection between the processes in nature not only in particular spheres but also in the inter-connection of these particular spheres themselves, and so can present in an approximately systematic form a comprehensive view of the interconnection in nature by means of the facts provided by empirical natural science itself.'

Marx-Engels: Selected Works, Vol. II, 1949, p. 352.

Introduction

Werner in the mining districts of Germany, and Hutton in the newly industrialized Scotland of the late eighteenth century, achieved its recognition as a practical science through the work of William Smith, the English surveyor and canal builder. Geology was indeed the dominant science in the somewhat repressed science in the decades that followed the fall of Napoleon. It had a solid and quiet appearance and it was believed that the study of the order of nature written in the rocks could only lend support to the Bible story of creation and through it to established religion and government.[1] This hope was to prove delusive and before the century was half over geology was the centre of controversy about the origins of strata, of hills and valleys and, through fossils, of the origin of species themselves. The formulation of Darwinian evolution followed as a logical if daring consequence of Lyell's *Principles of Geology*.[2] All through the century geology developed and spread mightily as the resources of the old and new worlds were first opened to exploration and use. It remained, however, largely descriptive. The physical and chemical bases of the phenomena of geology are even

[1] See C. C. Gillispie, *Genesis and Geology*, Oxford, 1951.

[2] The historical aspects of geology were early recognized. 'The history of the earth, and the history of the earth's inhabitants, as collected from phenomena, are governed by the same principles. Thus portions of knowledge which seek to travel back towards origin, whether of inert things or of the works of man, resemble each other. Both of them treat of events as connected by the thread of time and causation. In both we endeavour to learn accurately what the present is, and hence what the past has been. Both are historical sciences in the same sense.' W. Whewell, *History of the Inductive Sciences*, London, 1857, Vol. 3, p. 402.

now only beginning to be appreciated, in the measure that they are finding uses in scientific prospecting for oil and minerals.

The main lines of technological advance

The technical developments of the nineteenth century brought about a complete transformation of the manner of life of hundreds of millions of people in countries dominated by industrial production and mechanized agriculture, and notably affected the conditions of all the remaining population of the world.[1] Nevertheless, it is principally in *quantity* that the technical transformation of the nineteenth century was remarkable, in *quality* it was much less so. The basic invention of the Industrial Revolution—the use of power-driven machinery to take the place of handicraft—had already been achieved in the eighteenth century. What happened in the nineteenth was its enormous extension, together with a steady increase in its cheapness and efficiency.

At the outset the new technical developments were highly localized. The Industrial Revolution itself had been largely limited to Britain and even to a very small part of Britain, to the half-dozen areas where coal had been available as a cheap fuel. Though its products were spread throughout the world with unprecedented speed, the

[1] Out of the total estimated world population in 1901 of 1,608 million, some 290 million were to be found in the countries of Britain, France, Germany, Austro-Hungary, Italy, Holland and Belgium, all of which were markedly affected by industrialization, while 600 million lived under their direct colonial control and the remaining countries, including Russia and China, were commercially dominated by them.

methods of production, the pattern of the factory system, established itself but slowly in other areas. By the end of the century the only centres that rivalled Britain were those based on the coalfields of Pennsylvania and the Ruhr. Consequently, the pace of technical development in most traditional fields of industry was set by Britain. This remains broadly true even for the new chemical and electrical industries which later flourished in America and Germany because, at least in the nineteenth century, they still operated in a field dominated by the older industries.

To understand the pattern of technical advance it is therefore necessary, in the first place, to consider the conditions which determined it in Britain. In a bourgeois society—and Britain was the bourgeois society *par excellence* throughout the whole century—not only the day-to-day existence of technology, but its year-to-year change was determined by what profits it could show which in turn depended on the state of the market. Now the great feature of nineteenth-century Britain was the rapid growth of the market, however uneven that growth was. The very cheapness of even the earlier crude machine-made goods ensured a general increase in profitable production. Exports, first of consumption goods, largely textiles, and later of production goods, largely steel and machinery, went on increasing and brought with them a corresponding increase in imports of luxuries, of raw materials and finally of food for the ever-increasing industrial population.

The pattern of production was already set—it lay in exploiting the greatest early success of the Industrial Revolution—the new machine-made textiles. There seemed even little need to make radical improvements in spinning and weaving methods—merely to increase their scale was sufficient. The production of cotton cloth had increased from 40 million yards in 1785 to 6,500 million yards in 1887.[1] Such a phenomenal growth would by itself imply a proportionate increase in machinery, raw materials and ancillary processes such as bleaching and dyeing. Accompanied as it was by other developments in all fields of manufacturing industry it meant a scale of demand for raw materials and products that traditional methods of production could not supply and so still further forced the pace of mechanization. In themselves, the mechanical requirements of manufacture made little demand on science, though, as will be shown, the steam engine and the machine-building industry did so. It was rather the problem of providing the necessary, though relatively small-scale, ancillaries, particularly the chemical ones, such as dyes and soaps, to replace natural products, which had become too scarce and too dear, that provoked scientific solutions.[2] The chemical industry grew up largely under the shadow of the textile industry, while the gas industry found its earliest customers in the new mills and was later in return to provide the materials for the new coal-tar dyes. Machine production needed power;

[1] From then on the rate of increase of British—though not of world—production slackened, roughly comparable figures being 8,000 in 1912, the peak year, and then production itself actually fell to 3,600 in 1937 and 2,100 in 1950.

[2] An admirable and detailed account of how this occurred in Britain, and particularly in North Britain, in the eighteenth and early nineteenth century has recently been produced by Archibald and Nan Clow in *The Chemical Revolution*, London, 1952.

power meant steam, and steam called for coal and iron. The expansion of manufactures also placed a premium on rapid and cheap transportation and assured the success of the railways and the steamers.

The railway age

Cheap coal, the new universal fuel, and cheap iron, the new universal material, now both finally replacing wood in its cruder uses, are characteristic of nineteenth-century industry. The production of coal went up by fifteen times —from about 10 to 150 million tons—in the century between 1780 and 1880, a great advance though only a tenth as great as the increase in output of cloth. The increased output of iron, 110 times from 68,000 to 7,750,000 tons, however, almost kept pace with that of cloth in the same period. Both coal and iron were bulky goods needed in enormous quantities and the cost of moving them tied industry largely down to the coalfields. All through the century the steam engine and the factory chimney remained the symbols of the grimy, formless, gas-lit cities of the first industrial age. Nevertheless, it was this urgent need for heavy transport that gave rise to the greatest and most characteristic innovation of the nineteenth century—the railway.

The railway came straight out of the coalmines themselves, where rails had been used at least as far back as the fifteenth century. It was the convenience of running the trucks carrying the sea coal of Newcastle down to the staithes for loading on ships that led stage by stage to radical improvements. It was these improvements—first iron rails, then stationary engines, and finally locomotive

21

engines—that emancipated the railway from the mine and sent it, carrying goods and passengers, first over Britain and then over the world.

What the locomotive did for land transport, the steam-boat had already been doing for some years for the cheaper water transport, and with a far wider range, especially when, with the advent of steel, the ocean-going steamer was able to supplant the sailing ship. By these means the products of the new factories were enabled to undersell and ruin native industries all over the world, and to turn undeveloped countries into virtual or actual plantations supplying raw materials for industry. Towards the end of the century this type of trade was increasing, supplemented by an export of capital goods, largely rails and mining machinery, which facilitated the collection of agricultural products and ores over a still larger area, and was balanced by an increasing import of food.

The great transport revolution of the nineteenth century owed relatively little directly to science.[1] The first railway engineers were largely self-taught men, though few had as hard a struggle as George Stephenson. It did,

[1] Charles Babbage seems to have been one of the few who interested themselves scientifically in its working. In 1838 he fitted up—at his own cost, £300—a second-class carriage where he installed a self-recording apparatus measuring both the tractive force and the components of vibration of the carriage. With this he made extensive measurements—not without danger—for five months on the new Great Western line of whose broad gauge he was an enthusiastic supporter. In this example of operational research, as in his computing machine, he was a hundred years ahead of his time. See *Passages from the Life of a Philosopher*, London, 1864. Samuel Smiles and, curiously enough, Herbert Spencer were also connected with the early railways but they made no notable scientific contribution to them.

however, serve to stimulate and facilitate science in a number of different ways, particularly in geology and surveying at home and geography and biology abroad. Indeed, the great generalization of evolution was given its impetus largely as the result of the voyages of Darwin, Wallace and Huxley.

With transport went communication. The telegraph—the first practical application of the new electrical knowledge—followed the spread of the railways and was, within a few decades, to link the continents by submarine cables. Through these inventions the whole world could become one great market, in which the powers of business and finance could operate without the old limitations of space or time. Here, certainly, was a direct case of science leading industry—a small industry to be sure, but almost from the start a necessary one. And the electrical industry grew, rather more slowly than it might have done technically, as will be shown, but steadily. Towards the end of the century, the telegraph was to lead directly to the telephone, and at its very end to the wireless telegraph, the first opening of a new age of communication. The development of the telegraph provided a nursery for the young science of electromagnetism, supplying problems, part-time occupation, equipment and funds for the academic scientists and ensuring them plenty of students.

The telegraph and cable industries were also to be the source of the new electric light, traction and power industries of the eighties and nineties. This development, however, took the best part of half a century to realize and the reasons for the lag is one of the main questions that will be examined in the latter part of this book.

The metal industry

The great manufacturing and commercial effort of the nineteenth century could never have come into being or grown as it did, without a parallel and intense development of the old metal industries and the new industry of engineering. The needs of the new industry could not be satisfied in quantity or performance by the old universal material, wood. Wood was in fact used more than ever before, but it did not go far enough and it had not the strength for the new needs of machinery or structures. By the middle of the century cast and wrought iron had replaced wood for most machinery and was beginning to replace wood for ships and stone for bridges.

The basic inventions of the coal-fired blast furnace and of puddling and rolling had been made in the eighteenth century. They were not radically improved throughout most of the nineteenth with the exception of the introduction of hot blast by the gas engineer J. B. Neilson in 1828.[1] They were, however, enormously extended, in-

[1] This development was a striking early example of the direct influence of science on an industry normally resistant to it. J. B. Neilson was not an ironmaster himself but his father had been a works engineer at one of Dr. Roebuck's collieries. He became first foreman and then manager of the new Glasgow Gas Works and picked up the necessary chemistry at the Andersonian Institution, Glasgow. This first working-class technical institute was founded in 1793 through a legacy from Dr. John Anderson, the republican Professor of Natural Philosophy in the University. Its first principal was Dr. Birkbeck who was later to light the lamp of higher education in London (see p. 106 *n.*). In 1824 James Ewing, the ironmaster of Muirkirk, asked Neilson whether the variations in the yield of his furnaces could be suppressed by purifying the air used to blow them. Neilson had the ingenious idea of using the newly established principles of

creased in scale of operation and improved in hundreds of minor ways. These changes, moreover, were largely technical, the fruits of the experience of observant iron-workers and ironmasters. The subject was largely a closed book to science. It was very different with the great revolution in metal production that occurred in the second half of the century and ushered in the age of steel. This was largely due to the entry of science—amateur in the

expansion of gases to increase the volume, and hence the speed, of the blast by heating. After considerable obstruction he managed in 1828 to secure a full-scale trial of his idea. It was successful beyond all expectation but for quite other reasons than Neilson had imagined —effectively, by raising the combustion temperature and consequently saving nearly half the fuel used per ton of iron. (See Clow, *op. cit.*, and T. B. Mackenzie, *West of Scotland Iron and Steel Institute*, 1929.) This gave the Scottish iron industry a new lease of life so that it rapidly gained on the even more conservative English iron industry. This example is interesting in that it brings out a link between the gas industry and an important development in iron technology. That industry was itself a late by-product of the change-over from wood to coal technology. First coal was charred in open heaps to make coke as a substitute for charcoal in smelting iron. Then it was distilled in retorts for the tar as wood tar became too dear. Finally Murdock, an associate of Watt, thought of burning the previously wasted gaseous products. It was the first commercial realization of the scientific principle of gas collection that had started the great pneumatic revolution in chemistry. The Glasgow Gas Company was one of the most progressive in the world and Neilson had introduced notable improvements, particularly the old batswing burner whose yellow lambent flame still remains among our early memories. The early gas industry was an immensely profitable one, but to succeed it needed at the outset a blend of scientific research and practical capacity. Men like Neilson who had this experience were just the kind who could help to revolutionize the more traditional older industries.

case of Bessemer, more professional with Siemens and Gilchrist—into a traditional industry. For that reason, I have chosen it as the theme for a later chapter. The cheap steel that science thus brought in, a far stronger and more adaptable material than iron, provided and still provides the main material basis for a mechanical civilization.

Engineering

Throughout the nineteenth century the metal industry grew up with and fed an ever-expanding and diversifying machine-building and engineering industry. The nucleus of that industry had been formed in the eighteenth century through a blend of the traditions of heavy, and usually rough, construction of the smith and millwright with those of the fine workmanship of the clockmaker and instrument maker, itself an offshoot of the age-old goldsmiths' craft and allied with science since its inception in the Renaissance. Watt, trained as an instrument maker and turning himself into the first professional engine maker, may stand as the prototype of the new mathematical technical profession, while his partnership with the practical manufacturer Boulton in 1775 marked the foundation of the first engineering firm. The new industry had to find its own technicians and workers, who trained themselves as they produced the new machines. It was only by the eighteen-twenties that the next generation of men like George Stephenson, born and bred with engines, began to make themselves felt.

The aim of the industry was the production and multifarious use of power—the new power of the steam engine. The means to realize it were the development of strength

26

in materials, and design and precision in workmanship. In its pursuit of power the new engineering industry did not so much depend on science as create it. The successive developments of the steam engine—separate condenser, expansive working, compound cylinders—right down to the steam turbine at the end of the century were essentially successive essays in thermodynamics solved in practice before they were solved in theory. In Chapter II will be found a discussion of how the great generalizations of nineteenth-century physics, the conservation of energy and the limits of its utilization—the first and second laws of thermodynamics—arose out of attempts to understand and improve the performance of prime movers, predominantly of the steam engine.

On the side of metal working and of mechanisms, progress owed even less to science. The creators of modern machine tools—the great succession of Bramah, Maudslay, Whitworth, Roberts, Muir, Clements—all started as manual workers. They achieved, through an application of simple geometry and a deep experience of the behaviour of materials under stress, a steady improvement in accuracy and reproducibility of work. Not only did they learn how to work metal on a scale even larger than wood, but also they achieved a precision that wood could never have given.

The new accuracy was not merely pride in good craftsmanship. Left to itself that leads to beauty of form and ornament as is evident in the products of the seventeenth and eighteenth century. The working requirements of the new machines were for screws that would not work loose, for plane slides, for well-fitting pistons in accurately bored cylinders, and for wheels that must spin true. This imposed a new kind of craftsmanship, one where work was done from drawings, implying a deep understanding of three-dimensional geometry. The new engineer had in one way or another to acquire the rudiments of a scientific education. In his production fancy must be subservient to work and a new kind of beauty was created, one that appealed visually only to the initiated.

The new ability to machine metal accurately in turn made possible the series production of identical parts that could later be assembled. The impetus for this came first from America, where there was a shortage of skilled labour, with the elder Brunel, Eli Whitney and Colt. This method of manufacture started with small arms but was later to make possible the great labour-saving mechanisms of the sewing-machine, the typewriter, and the reaper and binder. But the full, logical development towards mass production by the adding of the mobile assembly line had, despite pioneer efforts in the pig slaughterhouses and in the box car assembly in America,[1] to wait till the twentieth century with the relatively enormous market offered by the motor-car.

The multiple machines of the nineteenth century, out-. side the pioneer industry of textiles, notably those in the printing trade, were the fruit of much ingenuity but relied little on science and gave little back to it. The typical inventor was usually a workman or amateur who contrived to find the most convenient arrangement of wheels, rollers, cogs and levers designed to imitate the

[1] See L. C. Ord, *Secrets of Industry*, London, 1945; S. Giedion, *Mechanization takes Command*, Oxford, 1948.

movement of the craftsman at higher speed and using steam power. The more flexible arrangements using electricity were scarcely available until well into the twentieth century.

Where the practical mechanics broke radically new ground was in the devising of machinery using forces greater than man could wield. Bramah's hydraulic press and Nasmyth's[1] steamhammer made heavy engineering

[1] James Nasmyth is an apparent exception to the rule that early nineteenth-century engineers started as mechanics. He was the son of a respected portrait painter of Edinburgh, himself descended through a long line of master builders from the eponymous Naesmyth who won his lands in the fight against the Douglas in the thirteenth century. His father, however, was familiar with many of the engineers of the day and young James showed an early passion for practical design and construction that won him instant employment as an assistant and not apprentice in Maudslay's shop.

How he came to invent the hammer is told by Nasmyth in his *Autobiography*, London, 1883, in Chapter XIII, curiously entitled 'My marriage—the steam hammer'. He was consulted in 1839 by Mr. Humphreys, the engineer of the Great Western Company, about the engines of a proposed paddlesteamer. 'I find', he wrote, 'that there is not a forge hammer in England or Scotland powerful enough to forge, the intermediate paddle shaft for the engines of the *Great Britain*! What am I to do? Do you think I might dare to use cast iron.' Nasmyth writes: 'This letter immediately set me athinking. . . . The obvious method was to contrive some method by which the ponderous block of iron should be lifted to a sufficient height—and then to let the block fall down on the forging. . . . I got out my Scheme Book . . . rapidly sketched out my Steam Hammer having it all clearly before me in my mind's eye.' After that Nasmyth did not make the first steamhammer; the paddle shaft of the *Great Britain* was never forged, the company decided to use screw propulsion— an example of one technical development cutting across another. Other opportunities did not arise. 'Very bad times for the iron trade

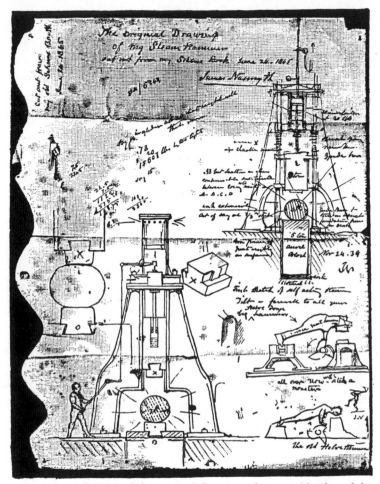

FIG. 1. Reproduction of the original drawing of Nasmyth's idea of the steamhammer. An amusing feature is the two small drawings in the bottom right-hand corner. Here we see 'J.N.' executing the personification of 'the old Helve Hammer' and the legend 'all over now with sitch a monster'.

possible, as the need for engines for steamers and later of guns and armour plate for battleships made it profitable. It was this development, which came in with cheap steel towards the last quarter of the century, that gave over-whelming advantages to the handful of big firms in Britain, Germany, France and America which alone could undertake this work and were the progenitors of the trusts and cartels of our time. They were even sufficiently wealthy to finance some research of their own, notably that of the steam turbine. Nevertheless, as will be shown in later chapters, the general tendency in engineering indus-try was to separate itself as an independent profession from the general trends of academic science. The eighteenth-and early nineteenth-century engineers, men like Smeaton, Watt and Rankine, were in the forefront of the science of their day; their successors, however eminent and suc-cessful as engineers, did not, with few exceptions like Sir Charles Parsons, contribute notably to the advance of science.

The chemical industry

It was not until the second half of the century, and only markedly so at its close, that a new and important industry —the heavy chemical industry—began to take its place

and for all mechanical undertakings set in about this time.' However Mr. Schneider of the Creusot works, itself set up by W. Wilkinson in 1782, on a visit to Nasmyth's establishment was shown the Scheme Book and he was to see his own hammer there in 1842. Nasmyth had no patent at this time—his partner thought the expenditure of so large a sum as £500 was unjustified—and all Nasmyth's money from his invention was sunk in the works. However he now secured one and got the full credit and profit.

beside the traditional industries. It was not based to the same extent as the electrical industry on academic science, but it tended to draw on it more and more as the century advanced, not only for improvements in technique but also for large-scale processes derived from laboratory experiments.

The chemical industry started and remained for most of the century largely ancillary to the textile industry.[1] It started from the need to supply acid, alkali and soap on a scale with which the old semi-domestic chemical industry could not cope. In 1785 the first great purely scientific addition was made with the introduction of bleaching by chlorine and bleaching powder. By the middle of the century it was adding dyes and the beginnings of plastics with mercerization and celluloid. The first big independent development came with the discovery of the explosive properties of nitrated cotton and glycerine. This was to lead to the creation of a new explosive industry, with dynamite and cordite appropriate to the new era of mining exploitation and for the wars that were soon to come.

Because of the close and evident links between the chemical industry and the advance of chemical science from the time of the great chemical revolution there is no need here to demonstrate their interdependence. The

[1] This is well brought out in the Clows' *The Chemical Revolution* who, in their successive chapters on kelp, soda, soap, sulphuric acid, bleaching, dyeing, and printing, show how the chemical industry, led by enthusiasts for the new rational chemistry of Black, Lavoisier, and Dalton, rose to the demands of the expansion due to mechanical spinning and weaving. Indeed if they had not done so the great textile boom which floated the Industrial Revolution could never have got under way.

problem is rather to unravel the closely intertwined strands of practical and theoretical considerations that led to the advance of both. The chemical factory could indeed never be far away from the laboratory where its processes had been first elaborated and in which they could be modified or superseded. In a later chapter an account will be given of examples of that interaction. It will suffice here to stress the fact that in this industry, no less than in electricity, there still remained, despite multiple links, a long lag between the laboratory and the works, and that, however scientific the principles used, the actual handling of materials in chemical factories often owed more to tradition and skill than to scientific design and operation.

Traditional methods could not cope with the enormous demands of the new industrial populations. Larger scale methods had to be found for brewing, distilling and preserving, and this opened a loophole for science, especially where new methods, such as canning and refrigeration, became technically possible. Some of the problems which affected the older science of physics and chemistry as well as the newly created science of bacteriology, are discussed in a later chapter.

Agriculture and medicine

It is not within the scope of this essay to discuss, except incidentally, the influence of the more purely biological industry—agriculture—on the progress[1] of science as a whole. Its effects were naturally most profound on the

[1] I hope to deal with some of these questions, notably the relations of agriculture and medicine to a rapidly expanding industrial community, in my forthcoming book *Science in History.*

development of biology itself, particularly in its contribution, through Darwin, to the great generalization of evolution. Nor is it within my province to discuss fully the mutual stimulation of the great art and profession of medicine and the sciences of human and animal biology, though the chemical controversy between Liebig and Pasteur discussed in Chapter III inevitably touches on this. I mention them here only to complete the picture of practical human activities, so as to avoid the impression that the external impetus to science in the nineteenth century came only from the predominantly physical and chemical industries.

Economic features of nineteenth-century industry

Taken in all, the enormous technical developments of the nineteenth century were aimed at achieving old-established ends by new means. To some extent this was a necessary transitional stage before the new forces, which the Industrial Revolution had liberated, could be sufficiently understood and mastered to enable new ends to be envisaged and achieved. To an even greater extent, however, it was due to the dominance throughout the whole century of a market economy concerned essentially with producing the goods that were in known demand but producing them in far larger quantities and making a far larger profit than the old handicraft industries could manage. The advantages of enormously more efficient methods of production did not accrue to the great bulk of the population of the industrial countries and still less to those in their colonial dependencies. It was largely divested from them by the operation of an unjust, unstable and wasteful economic system. The new goods involved the use of far less labour per unit of manufacture but the ingenuity that had been devoted to production had concentrated on cheapness rather than quality or serviceability. Still less was there, or could there be, an attempt to meet the real need of people by designing these objects or devices for this purpose—they were produced for profit rather than use.

Production for profit was the great motive force that had stimulated the Industrial Revolution into being. The force that continually modified production was the need to use part of that profit as further capital investment. The pressure of competition did lead to more efficient and therefore more profitable techniques—though in a very irregular way, as will be shown in later chapters—and it was here the services of science were mostly in demand. It would be more true to say that it was here that science could insinuate itself, for one leading characteristic of the chase for profit was its blindness—a blindness not concealed but even praised by the dominant advocates of *laisser-faire*.

Though free competition was still the watchword, in the latter part of the century it tended to give way to price rings, cartels and monopolies. The early successes of the new industries were making capital harder and harder to invest profitably at home. Alternative investments were found in the opening up of colonial territories and in the war preparations due to the resulting rivalries between the great powers. These could only be negotiated in larger units than the old individual entrepreneur or family firm could command. Large-scale monopoly industries did not achieve their full development till the twentieth century, but already before the end of the nineteenth the most rapidly advancing fields of technique, and consequently of science, were those in the purview of the great telegraph companies, chemical companies, steel companies and armament firms. The era of the industrial and governmental research laboratories, where science was organized and directed to particular ends, was just about to begin.

Relations between science and industry

The broad outlines of the development of science and of industry throughout the nineteenth century already show something of the character of the most general connections between them. Two main complementary trends are evident. On the one hand we have the scientific study of already established industrial processes, such as the use of steam engines or the making of iron, which lead to new scientific generalizations, such as the conservation of energy or radiation physics;[1] on the other we have scientific discoveries, particularly in electricity and chemistry, that give rise to new industries such as the telegraph and synthetic dyes. These relations, now easy to recognize, were not so easily grasped at the time. Despite much lip service to the mutual dependence of science and industry the anarchic development of industry made it virtually impossible for the links between science and industry to have any rational or planned basis. Indeed,

[1] The relation of the steam engine to the laws of energy are later discussed in this essay. Compared to that, the development of the radiation pyrometer as a means of assessing furnace temperature, represents a minor connection between industry and science. Yet its results were to lead through the laws of energy distribution in radiation to Planck's revolutionary generalization.

as will be shown, they were of an even more casual and individual character than those between the various branches of industry. While it was generally admitted that the age of science had arrived, little or no provision was made until the very end of the century for the development of science itself, let alone its application to practical purposes. Nevertheless it was possible for individual scientists, alone or in conjunction with far-sighted business men, to study some of the problems of industry and to propose solutions, some of which, though usually after a considerable lapse of time, became embodied in practice.

A serious, detailed study of the relations of industry and science in the nineteenth century would be a task of great interest and no little profit, for it would certainly reveal many abandoned enterprises which have now become feasible as a result of advances in other fields. It would, however, be a very hard task, not because the material is not available, but because of its very copiousness and confusion. What are needed are statistical and quantitative surveys of the developments in each industry and of the relations between them. We should want to know the scientific capacities of those directing industry on one hand, and on the other, the circumstances and incentives that attracted scientific men to industrial problems. Only on the basis of the collated study of such cases could a reliable general picture be formed. Here, in this brief and merely exploratory study, it is beyond my powers to attempt anything of the sort. The best I have been able to do is to choose four examples, illustrative of aspects of the interaction of the physical sciences with industry. These form the subject of the following four chapters of this essay. They are:

Chapter II. *Heat and Energy*, being a description of the discovery of the laws of transformation of energy known as the first and second laws of thermodynamics.

Chapter III. *Ferments and Microbes*, being an account of the controversies of Liebig and Pasteur and of the links between organic chemistry, agriculture and medicine.

Chapter IV. *The Age of Steel*, in which the contributions of Bessemer, Siemens, and Gilchrist Thomas are discussed.

Chapter V. *Electric Light and Power*, centring round the work of Swan and Edison.

These subjects have been chosen with the dual purpose of covering the fields of physical and chemical science and of technology and also of presenting some picture of the changes in the relations between science and techniques that occurred in the course of the century. Several other similar themes suggest themselves, such as the history of the telegraph, of refrigeration or of the introduction of aniline dyes or high explosives, but it should be possible to see, even in the restricted selection given here, and without any attempt to provide a detailed historical account, something of the general mechanism of interaction.

Chapter II

HEAT AND ENERGY

THE central and most far-reaching discovery in the physical science of the nineteenth century was that of the conservation of energy—the first law of thermodynamics. Inseparable from it is the second law of thermodynamics, which sets a limit to the amount of energy available for mechanical work and which may accordingly be called the principle of the dissipation of energy.

By the time these laws come fully into the light of scientific history in the middle of the century they are associated with the names of many men, some already important figures, like Rankine, Thomson, Joule and Helmholtz, some now almost forgotten, like Séguin, Holzmann, Hirn and Colding, and the solitary and tragic figure of J. R. Mayer. Indeed over the whole question of the first discovery of the conservation of energy there raged a bitter controversy in which scientific reputations were endangered; great scientists like Thomson and Tyndall abused each other in public,[1] teachers like Dühring

[1] See pp. 64 ff. *n.*

39

lost their posts, and poor Mayer was rewarded for his epoch-making discovery by slander and ridicule which drove him to attempted suicide and nearly to the madhouse.

It should be clear from the mere existence of this crowd of claimants and the fact that the idea of the conservation of energy was picked up along several different lines of approach that it must have been dead ripe by the forties. As we shall see, like other epochal discoveries it was missed not because it was concealed in natural phenomena but because, for traditional and academic reasons, it was unlikely to be looked for and hard to recognize. Though obvious and ultimately obtrusive, it did not fit into any then recognized scheme of knowledge and did not follow from generally accepted principles.

Only a mind at once concerned with the problem of the practical production of energy, free from preconceptions, with a clear vision and physical competence, could have been expected to discover it before its term. Such a man did exist, one of the great lost geniuses of all time, the young engineer Sadi Carnot, son of the 'organiser of victory' of the French Revolution, Lazare Carnot. In 1824 he published the treatise, *Réflexions sur la puissance motrice du feu et sur les machines propre à developper cette puissance*, in which he enunciated the second law of thermodynamics in the form of a theoretical limit to the efficiency of engines reached through a *reversible* cycle of operations. We know now[1] that he also discovered the first law and found a value for the mechanical equivalent

---

[1] His notes were found in manuscript long after his death and only published in 1878. His enunciation of the principle of the conservation of energy (*puissance motrice*) is as clear as any given later.

of heat in 1830. He never published it for he was to die from cholera in 1832. If he had lived and if, though it would have been no easy matter, he had been able to impress his contemporaries with the importance of his work, the science of thermodynamics would have been advanced by at least ten years.

'La chaleur n'est autre chose que la puissance motrice ou plutôt que le mouvement qui a changé de forme, c'est un mouvement. Partout où il y a destruction de puissance motrice dans les particles des corps, il y a en même temps production de chaleur en quantité précisement proportionnelle à la quantité de puissance motrice détruite; reciproquement, partout ou il y a destruction de chaleur, il y a production de puissance motrice.

'On peut donc poser en thèse générale que la puissance motrice est en quantité invariable dans la nature, qu'elle n'est jamais à proprement parler ni produite, ni détruite. A la vérité elle change de forme, c'est-a-dire qu'elle produit tantôt un genre de mouvement, tantôt un autre, mais elle n'est jamais anéantie.'

Sadi Carnot: Biographie et Manuscrit,
Academie des Sciences, Paris, 1927, p. 81.

('Heat is nothing but motive power (energy) or rather it is motion that has changed its form, it is a (form of) motion. Wherever there is destruction of motive power in the particles of bodies, there is at the same time a production of heat in a quantity precisely proportional to the quantity of motive power destroyed; reciprocally, wherever there is destruction of heat, there is a production of motive power.

'One can thus propound the general thesis that motive power (energy) is an invariable quantity in nature, that it is never, properly speaking, either produced or destroyed. In fact it changes its form, that is to say it produces sometimes one kind of motion, sometimes another, but it is never destroyed.')

It is peculiarly tragic that the scientific importance of this great principle was entirely overlooked by the brother—himself a distinguished scientist—into whose hands these notes passed.

Roads to the conservation of energy

But whether the solution was to be reached earlier, by individual intuitions and faultless logic, or later, by patient experiment and the effect of accumulated evidence of many workers, matters little in the general course of the history of science. There were, in fact, a sufficient number of channels of interest which were all converging on the same junction to make the ultimate discovery inevitable. The first and most urgent of these was the technical concern in the economic performance of the steam engine. Originally the steam engine had been 'a means of raising water by fire' and the question of how much fire could raise how much water was bound to be put once the wonder of the original achievement had passed. Indeed it had already led in the eighteenth century to Watt's criticism and improvement of the old atmospheric Newcomen engine, which was the first technical application of the scientific study of latent heat by Black, with whose experiments Watt had been closely associated. The science of the steam engine was thus linked with the late eighteenth- and early nineteenth-century pneumatic interest, particularly with the study of the thermal expansion and specific heats of gases by Dalton and the French physicists, Charles, Gay-Lussac and Regnault. Meanwhile, though steam engines were being continually improved by hunches and by trial and error, nothing was known as to what limited their *duty* (water raised per pound of coal burnt) or how near that duty approached any ideal limit of performance. As Engels said 'the practical mechanics of the engineer arrives at the concept of *work* and forces it on the theoreticians'.[1]

[1] *Dialectics of Nature*, London, 1940, p. 75.

The second channel of interest was the chemical and biological one drawing mainly from Lavoisier's masterly and quantitative explanation of the origin of *animal heat* in the combustion of nutriment. Animal heat, previously a mysterious entity in its own right and supposed to be resident in the heart,[1] was shown to be the same in essence as any heat liberated by chemical change. This, however, left two further questions unanswered: how does chemical change produce heat and what is the source of animal power? This channel, though less technical than the first, was the way in which the medical doctors, Mayer and Helmholtz, were led to the problem of the conservation of energy.

The third channel was the appreciation of the new powers being revealed by magnetic and electrical phenomena, beginning in 1821 with Oersted's discovery of the magnetic effects of currents and leading to the development of the electromagnet and the electric motor by Sturgeon in England, Henry in America and Jacobi in Russia. When the reverse transformation of magnetism into electricity was discovered by Faraday in 1831 it was apparent that a new source of power was available, but it remained to be discovered on what terms it could be exchanged with the older forms. This was the problem that was to lead young Joule, between 1840 and 1847, into the most precise determination of the quantitative relations of electricity, heat and mechanical work.

These considerations show in how many ways the needs of an expanding economy imposed the problem of an

[1] See Sir C. Sherrington, *The Endeavour of Jean Fernel*, Cambridge, 1946.

effective accountancy for forms of energy, so that all could be equated with the universal standard—money.[1] The solution of this problem required not only precise measurements on a reasonable scale but a physical and mathematical theory which would make the experimental results intelligible.

Now both of these were well within the range of eighteenth-century science. Rumford's experiments, which pointed the way to the production of heat by work, were on a larger scale than any carried out by Joule, while Cavendish could have readily provided the required accuracy. On the theoretical side the idea of heat as an expression of molecular motion had already been put forward by Bernoulli, Euler and Lomonosov and the necessary mathematics for expressing thermodynamic relations was available in the generalized mechanics of D'Alembert and Lagrange, all in the eighteenth century.

[1] The history of the accountancy of energy in economic terms is a long one. It begins effectively with Watt's ingenious method of charging royalty on the mine pumping engines he set up in Cornwall under his much maligned patent of 1775. Here, as the charge was based on the difference between the cost of engine coals and horse feed for the same 'duty' of pumping, Watt had to establish the 'power' or rate of doing work of an ideal horse at 33,000 foot-pounds per minute, a unit which we still use, though its technical and economic origin is forgotten. Later when engines were bought and not erected by the makers, *power* and *duty* had a technical rather than economic importance, but the search for minimum fuel for maximum work went on. The first power that could be sold on a large scale was electrical. Faraday's reply to Gladstone, who asked what use was electricity, was: 'Some day you will tax it.' From the moment central electrical stations were set up energy was sold as such, though the appropriate unit, the watt, did not receive its name till 1882.

Something more was needed before the theory could come fully to light and that missing factor was the combination of the mathematician and the practical engineer or experimenter, either in one man as with Carnot, or by close collaboration such as that of Joule and Thomson. It was thus necessary to wait for the training of a sufficient number of engineers familiar with the steam engine, which in turn depended on the phenomenally rapid increase in its use in the early nineteenth century. To a lesser degree it depended on a sound mathematical training for engineers such as was first provided by the École Polytechnique in Paris, one of the direct products of the French Revolution. The École and the steam engine led to Carnot's great first step, but the steam engine was still too little known in France, or perhaps not considered scientifically respectable enough, for his idea to take roo in its native country. When his ideas reappeared, twenty years later, they were in the form of theoretical formulation in Germany and of experimental verification in England.

These general considerations can be reinforced by a detailed study of the actual ways in which the different pioneers of the laws of thermodynamics arrived at their results and of the efforts they had subsequently in getting them accepted. There is no space here to describe them in detail, but some selected notes may help to make the process clear.

Sadi Carnot and the principle of reversibility

Of all the pioneers Carnot is far the clearest, the most logical, and shows the greatest physical insight. In his time there was a great inventive rush of new engine designs in Britain where high-pressure, multiple expansion, vapour and air engines were being tried out and where exaggerated claims were being made for their efficiency. Carnot standing a little aside, though fully aware as his book shows of the details of British practice, constructed in his mind his ideal engine, one that, working with any liquid or vapour, could be used to give the best possible result. From the engineering point of view it was a generalization of the expansive working of steam which Watt had introduced and Hornblower and Woolf used in their compound engines. Theoretically Carnot started with the accepted conception of *caloric*—the elastic, imponderable fluid of Lavoisier and Dalton.[1] He pictured the engine as working by taking caloric from the boiler or source and throwing it out in the condenser or sink, just as a mill uses water at the headrace and lets it run out at the mill tail. He saw something else from this analogy, that in order to get mechanical work the equilibrium of caloric must be disturbed so that caloric must also have an intensity, head or potential as well as a quantity. This was to lead him, too late for posterity for the

[1] The word caloric was introduced by the Lavoisier school in their revision of chemical terminology (*Méthode de Nomenclature Chimique*, 1787). It took the place of Black's 'matter of heat' but retained, in spite of the inconclusiveness of all attempts to find its weight, the character of a material fluid (see McKie and Heathcote, *The Discovery of Specific and Latent Heats*, London, 1935). In itself this did not seem so unreasonable in view of the existence of another elusive imponderable fluid, electricity. Nevertheless, it was deemed by all but the acutest minds to preclude the picture of heat as a form of motion.

knowledge was to die with him, to the concept of the equivalence of work and heat.[1] What he had called caloric and claimed was conserved was what was later called *entropy* $\frac{Q}{T}$ which is, in fact, conserved in his hypothetical reversible system.

This concept of ideal reversible working was indeed the most profound and far-reaching of Carnot's ideas. He had understood the essential difference between the steady and *irreversible* flow of heat from a hot body to a cold where, in his language, caloric fell in a way from which no work could be extracted, and the reversible expansions and compressions where, again in his language, caloric could fall and generate motive power or motive power could be absorbed in raising caloric. An engine working on a completely reversible cycle between two temperatures must be the most efficient, for if there was one more efficient, it could drive the former in reverse and so produce perpetual motion. It is worth noting that Carnot's last work proves that the first as well as the second law of thermodynamics could be deduced entirely from consideration of the performance of the steam engine. Carnot had fully grasped the essential factor which makes any

[1] Carnot does give, in his published work *Réflexions sur la puissance motrice du feu*, though this seems to have escaped notice then or later, a measure of the mechanical equivalent of heat by considering an engine working reversibly between 1000°C. and 0°C. where the theoretical efficiency is 80 per cent and gets the value 560 kg. metres per calorie against the now accepted value of 427. By 1830 we know from his unpublished notes he had got the value of 370, very near the 365 of Mayer twelve years later, and probably by the same method. On this reckoning he finds the engines of his day to be less than 1 per cent efficient.

heat engine efficient, a large fall in temperature of the working fluid. In a footnote on p. 110 he advocates for this reason an internal combustion engine with pre-compression[1] and even suggests the use of exhaust gases for raising steam.

[1] This was based on the scientifically interesting but economically useless machine of Niepce, the pioneer of photography, which used Lycopodium powder as fuel inside the cylinder. The economic value of such an engine was beginning to be perceived about this time. In an article on 'Gas Power' by B. Cheverton in 1826, quoted in E. Galloway's *History of the Steam Engine*, London, 1830, we read:

'It has long been a desideratum in practical mechanics (Mr. Cheverton observes), to possess a power engine which shall be ready for use at any time, capable of being put in motion without any extra consumption of means, and without loss of time in its preparation. These qualities would make it applicable in cases where but a small power is wanted, and only occasionally required. They are so numerous, and the consequent saving of human strength would be so great, that the advantages accruing to society would be immense if even the current expense were much greater than that of steam. Such an engine should also be actuated by a force so concentrated, and so compendiously appropriated, as to occupy but little space, and be but of little weight, by which it would become applicable to locomotive purposes. If, in addition to this, the consumption of materials was moderate, we should then be in possession of a mechanical agent, which would enable us to navigate the *ocean* independently of the wind, but which it is in vain to expect from our present means. It is well known that the common steam engine satisfies none of these conditions' (p. 659).

It is interesting to note that, though the need and the principles of internal combustion engines were understood, another fifty years of desultory development were needed before they were being produced commercially on a large scale and another forty before the revolution which Mr. Cheverton anticipated came to pass.

Carnot's treatise found few readers, its full scope was not realized for twenty-five years, if then, but it was read by at least one competent person, the engineer Clapeyron, who used it as the basis of a paper which in turn attracted the attention of Thomson[1] in Scotland and Clausius in Germany. This tenuous thread, however, was sufficient to secure that the main content of Carnot's work was never lost. It was, however, distorted by Clapeyron through his rigorous interpretation of the conservation of caloric. It was adherence to this idea that prevented Thomson, until 1851, and with him the learned world, from accepting the overwhelming evidence of Joule's experimental work on the conversion of mechanical work into heat.

Energy and the locomotive. George Stephenson

This conception, however, was being more and more forced on engineers by the change from the stationary and generally low-pressure engines of the eighteenth century to the new high-pressure engines introduced by Trevithick in 1800 but finding little use until the railways gave them their chance in 1829. A pumping engine could afford to work slowly. It might last a hundred years, so that its prime cost mattered little and economy of fuel was the prime consideration. The same was largely true for the steamboat engine. With the locomotive engine it was different. Limited as it was in size, it had to produce

---

[1] Thomson never saw the original *Puissance Motrice* till 1848; he failed to find a single copy during his stay in Paris in 1845. If he had, he would have almost certainly been able to reconcile himself to Joule's ideas far earlier.

enough power to draw trains perceptibly faster than horses could. Far less science had gone into locomotives than stationary engines. George Stephenson, a self-taught engineman's son, was a great contrast to the learned Watt of fifty years before. Stephenson got the power for his 'Rocket', which won him pre-eminence, by the simple devices of the tubular boiler and the draught from the cylinder exhaust steam jet, which he turned up the funnel. This was not done for any scientific reason but in the first place because the whistling steam as it left the cylinder frightened the horses on the road as it wheezed along at four miles an hour. By converting his engine into a buffer he so increased its speed to a vertiginous twelve miles an hour that it became more terrifying than ever. Stephenson's genius went much further than that. The work that made him pre-eminent in locomotive design and manufacture was his grasp of the unity of vehicle and road— 'like man and wife' as he put it. The first locomotives had to perform on tracks neither level nor even and it was only by his superb combination of practical experience and real grasp of scientific principles that Stephenson made them work and pay where so many others, including the unfortunate genius Trevithick, had failed. His first engine, completed in 1814 and soon appropriately named 'Blutcher', had no springs and practically jolted itself to pieces. He brought out another model a year later where this defect was remedied with a boiler supported on vertical steam-filled cylinders, the first pneumatic shock absorbers, see Fig. 2 *a*. These were the ancestors of the famous 'Rocket', Fig. 2 *b*, which carried off the palm at the Rainhill trials in 1829 on the Manchester and Liver-

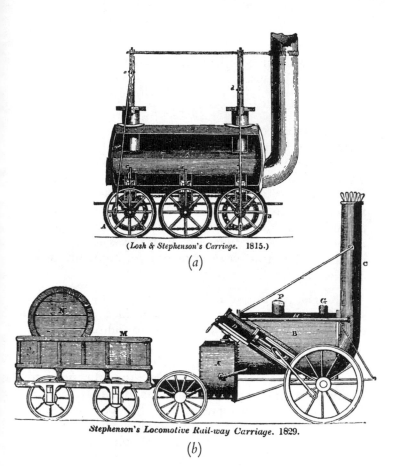

(*Losh & Stephenson's Carriage.* 1815.)

(*a*)

Stephenson's Locomotive Rail-way Carriage. 1829.

(*b*)

FIG. 2. Two stages in Stephenson's development of the locomotive. The original carriage with its tubular boiler was used for hauling coal wagons at a maximum speed of 4 m.p.h. The second, the famous 'Rocket', with the vital improvement of the steam-forced draught was capable of a speed of 30 m.p.h. and easily distanced its contemporary competitors.

Illustration from *The History and Progress of the Steam Engine,*
Elijah Galloway, 1830.

pool railway which he had surveyed and built together with his son Robert.[1]

Séguin

It was in the locomotive, as Mayer said, that heat is distilled out of the boiler, turned into mechanical work in the moving wheels, and condensed again to heat in the axles, tyres and rails. The first appreciation of this came from one of the pioneers of railway construction in France, Marc Séguin. He was one of the most conscious enthusiasts for the Industrial Revolution and his introduction to *Traité sur l'Influence des Chemins de Fer* (1839) is perhaps the most lyrical panegyric of capitalism in its creative stage:

'To increase the well-being and the enjoyment of material life is today the dominant idea of civilized nations. All their efforts are turned to industry because it is from that alone that one can expect progress. It is industry that gives birth to and develops in mankind new needs and gives them at the same time the means to satisfy them. Industry has become the life of the peoples. It is thus to this development that all hopes and wishes should tend, all our talents and knowledge be devoted. It is around this powerful lever that should be united all superior spirits that aspire to the honour of contributing to our social regeneration.

[1] It is interesting to note the pains that George Stephenson took to ensure that his son had the mathematical and scientific education which he had missed and what good use he made of it. See S. Smiles, *Life of George and Robert Stephenson*.

Heat and Energy

'Where are the limits before which human power will come to a stop? Commonplace individuals can never imagine them beyond their own horizon but nevertheless every day that horizon is widened. Every day its limits are put back. Look around us, everywhere in the last twenty years the elements of the old civilization have been modified, perfected, renewed; everywhere wonders have been worked. The enjoyment, the commodity of life which had been reserved only for men of fortune are now enjoyed by artisans. In a few more steps they will be shared equally with all classes. A thousand inventions are simultaneously born and lead to other discoveries and these in turn will become the starting-point for new progress; all these changes concur to the profit of the whole public and tend to make well-being common property. This is a new era based on the love of the good and the beautiful which is being built on the ruins of class prejudices and the monopolies of the wealthy. In all its creation, in all its innovations the same character is shown. Low prices and general utility are the essential conditions of the vitality of the industrial arts. Governments, national as well as local, drawn into this irresistible mass movement have followed the same drive and it is only by granting reforms to meet the needs of the times that they have maintained themselves in office, submitting these reforms to the point of view of modern ideas that they have been able to secure their acceptance. The old world has shaken off the yoke of its old habits. It is refreshing and remaking itself. So look, everything is changed around us—the towns, the face of the countryside, the course of the rivers, the work of the peoples, the production of the soil and industry, the dis-

tribution of property—everything has taken a new face.
And just where the direct force of the material power of
men has shown itself insufficient to accomplish its work
and to persevere in progress; where his will seems to be
broken against insurmountable obstacles, just there a drop
of water turned into steam acts to supplement his weakness,
to create for him a power of which we cannot now, nor
yet for a long time to come, measure the extent.

'From now on with the help of this agent these prodigies
have been accomplished, and the wonders which our
fathers would not have thought realizable with the united
efforts of all their magicians have become the ordinary
run of things. Machines which only need from man an
idle supervision now spin and weave by themselves linen,
cotton, wool and silk and return to us in a variety of
stuffs the material which we supply to them in the raw
state. Then after having been submitted to a chemical
preparation for some hours these cloths dipped into a vat
come out as if by enchantment with the most lovely
colours, with the most graceful designs. These make the
pretty prints, which on holiday, the working class itself
wears, and which in the country as much as the town,
light up with their sparkle and freshness groups of young
girls and spread all around them an atmosphere of joy,
comfort and happiness. Besides, the old rags that you
throw into the masher soon come back transformed into
paper of the purest whiteness ready to receive and spread
and make eternal your thoughts; some minutes only are
sufficient to achieve this metamorphosis. Everywhere the
most delicate objects of use and of luxury are poured into
consumption at prices which are always decreasing. By

means of that same steam, rivers, seas are navigated. It transports us with inconceivable speed to all the extremities of the world in floating palaces which shelter the poor man, the rich man, offering them alike luxury and conveniences which they often lack in their own houses. Finally, in our own valleys and across our hills wind and spread long ribbons of iron, along which rush, rapid as thought, those formidable machines which seem to eat up space with a spontaneous impatience and which seem almost alive in their breathing and their movement. When we consider the majestic elegance of these lines which speed with grace and level themselves out to cross plains, valleys, precipices and the granite mountains; when one hears the noise of these trains which carry thousands of individuals so fast that you cannot distinguish them; when one remembers that these results are the work of an industry which is only a few years old, of an agent which we have as yet studied only very imperfectly, an art which is in its infancy, one asks what will be the last prodigies to be realized by the perfectioning of this art, and one feels a noble desire to contribute to the realization of these incalculable blessings.'

This lyrical appreciation of the age of steam is doubly interesting as coming, not from a young enthusiast, but from a mature man who still carried into the nineteenth century the spirit of the eighteenth-century philosophers. It was too early, certainly in France of 1839, to see the reverse side of the medal—only some 2,000 miles of rail had been laid there by 1850. Many other idealists saw in the railways the road to the millennium. A substantial proportion of the first French railway-builders were fervent disciples of Saint-Simon.

Séguin's approach to the problem of heat arose as he expressed it himself out of the economics of railway practice:

'To consider the present state of the railway industry; to indicate the points where it seems capable of improvement; to call attention of science to the gaps which have to be filled; finally to express some personal ideas which perhaps will be of use in the future; that is the aim which I propose myself in publishing this book. I have no illusion, for that matter, of the fate which is reserved for it. I know that in dealing with an industry that is only born yesterday, where progress of the day before is always wiped out by that of tomorrow, the observations and the ideas which it contains will be outstripped; but far from drawing back before such an anticipation I have accepted it with hope. My deepest wish is to see these pages quickly abandoned by the practical men and relegated to the back of the libraries, as documents which are now only of historical interest.

'Devoted to industry since my youth, I have occupied myself before everything in improving the system of communications in France. Some journeys in England have convinced me that for bringing to my country the industrial civilization of the English nation, it would be necessary, before everything, to put our means of transport on the same level as theirs. For that reason we must multiply our bridges, activate our steam navigation and establish railways. It was towards the accomplishment of this task that I directed my efforts. In 1824 I constructed the first iron wire suspension bridge over a large river. The eagerness which was shown on all sides to imitate this example soon surpassed all my hopes. The simplicity, the elegance and above all the cheapness of these bridges recommend them to public favour, and in a few years a great number of them were put up in localities where arched bridges would have been impossible or too expensive. The application of steam to navigation and to railways presents many more difficulties and it is principally the story of the efforts I have made to improve the system of machines that I wish to put now before the public.'

He clearly states the equivalence of heat and work, attributing it to his uncle, the famous Montgolfier, who, in 1783, had invented the fire balloon, itself a type of heat engine.[1] He obtained a numerical estimate of the right order of magnitude for the mechanical equivalent of heat, based on the degree of cooling of steam by expansive working. As, however, most of his treatise is concerned strictly with railway engineering it is not surprising that it attracted little attention, especially from a scientific world that was affecting to scorn practical utility.

The line of approach which was ultimately to bring the conservation of energy into the full light of science was not to be the engineers'. They were concerned with turning heat into energy, but inevitably at the start with such low and variable efficiency that, even if such men as Carnot, Séguin and later Rankine could penetrate the

---

[1] Montgolfier was also a pioneer of the internal combustion engine. His 'pyrobelier' was a century ahead of its time.

logic of its working, the precise equivalence of work and heat was far to seek. It was otherwise with the inverse process. The production of heat or fire by work, striking with a flint or rubbing with wood, goes back to the very origin of civilization and is enshrined in the rituals of sacred flints and fire-making. There was an uninterrupted tradition through the ancients, elaborated by seventeenth-century scientists such as Hooke, that heat was a form of motion. The experiments of Rumford and Davy, qualitative and fallacious as they may have been, confirmed it. However, the belief in caloric, with the authority of Lavoisier behind it,[1] prevented them being followed up for forty years. After that the current was too strong.

Mayer and Joule

The two men who did most to break through the official silence were Julius Robert Mayer and James Prescott Joule. Their approach was diametrically different: Mayer conceived the idea and did experiments to support it; Joule did experiments which in the end convinced him of the equivalence of work and heat.

[1] Lavoisier considered caloric—a weightless fluid—a necessity to balance his chemical equations. In effect, this amounted to expressing chemical changes in the form of equations equivalent to:

$$\text{Hydrogen} + \text{Oxygen} = \text{Water} + \text{Caloric}$$

All the weight of the oxygen and hydrogen was to be found in the water but in coming together they were deemed to squeeze out a definite and measurable quantity of caloric—the heat of reaction. These quantities were usually so great as to completely mask any chemical energy that was turned into mechanical work. Lavoisier's idea coupled with Dalton's atomism contained the germ of the quantum theory of chemistry of 130 years later. In the interval however, by dignifying caloric, it acted as a bar on scientific advance.

Heat and Energy

In the case of Mayer, a philosophically inclined young doctor, it was the leisure of a long sea voyage, helped on by hints about the redder blood of sailors in the tropics, where they needed less energy, and a helmsman's yarn about water being warmer in a storm, that set him thinking about the equivalence of heat and motion. His leading idea was that there was an analogy between the motion produced by bodies falling together under gravity and the heat produced when a gas was compressed. When he got back to his native Heilbronn in 1841 he sent a paper containing these ideas to the leading physical journal, *Poggendorf's Annalen*. It was not refused, it was simply left unacknowledged, to be discovered forty years later. Mayer was deeply disappointed, but on the advice of a friend he attempted to back his theoretical ideas by experiment. He showed that water could be heated by shaking but, as with virtually no apparatus he could not determine how much work was lost, he used data already accumulated on the specific heats of air at constant volume and constant pressure to calculate the heat produced by a definite amount of work in compressing a gas. This paper he sent to Liebig's *Annalen der Chimie* and Liebig was enterprising enough to publish it in 1842, thus definitely establishing Mayer's priority. Nevertheless by being published in a chemical journal it escaped notice of the physicists and engineers, particularly in Britain, who could have used it, so that the third approach to the formulation of the conservation of energy, and ultimately the one that was to secure acceptance in the learned world, was made quite independently by Joule.

Joule, a wealthy young brewer's son, inspired by an

early training in science by John Dalton, had ample leisure and money to do what experiments he liked. Being brought up in Manchester at the height of its activity, he naturally turned to questions of the usefulness of physical forces, particularly the then fashionable force of electricity. His first experiments showed that the early hopes of obtaining infinite quantities of work from the new electromagnetic motors were illusory. He showed that all the energy came from using up the zinc in the battery. His papers on this received little attention largely because they lay altogether outside the predominantly Newtonian ideas of the British academic world.[1] His later experiments

[1] Osborne Reynolds points out that the failure of the Royal Society to publish Joule's 1843 paper, except in abstract, was useful to him if not to science. If it had been published, he thinks, others more able or more eminent would have seized on the idea and Joule would have had no chance to develop it. (*Memoirs and Proceedings of the Manchester Literary and Philosophical Society*, Vol. 6, 4th Series, 1892, p. 50.)

Professor Rosenfeld in his penetrating account of Joule's ideas (*Bull. Soc. Hist. Sci.*, Vol. 1, 1952, No. 7, p. 171), which also comments very favourably on Mayer's contribution to the discovery of the conservation of energy, remarks:

'The socially significant circumstance in Joule's attitude to a purely scientific problem is, however, the direction that his curiosity took. The attention that he paid to the relations of equivalence between heat and other physical phenomena was an entirely novel outlook at the time. It was a kind of curiosity quite different from that which inspired Davy and Faraday in looking at the same phenomenon. In fact, we have here one of the most striking examples of a physical phenomenon looked upon by great physicists from two quite different points of view. Joule's way of looking at it was so strange, so novel, that it was not immediately understood, even by the most prominent scientists of the time. This is convincingly demonstrated

culminated in his measurement of the heat produced by a known amount of work by agitating water. They were primarily undertaken not so much to add to his own fixed conviction of the equivalence of heat and work,

by the fact that the first paper submitted by Joule to the Royal Society was not judged of sufficient interest to be published in full.

'. . . It was only when the accumulation of wealth in the closing years of the eighteenth century created a demand for increased production that the need was felt of replacing hand tools by machine tools, and then of supplying power to drive the machines. These changes in the methods of production brought about a revolution in the social set-up of society. A new class of people arose whose activity centred on manufacturing rather than on commerce. The interest of those people in science was naturally directed towards the utilization of the forces of Nature in obtaining mechanical power. It is in just such a community that Joule grew up. The son of a wealthy brewer, he could look forward to a life of leisure. Thus, on the one hand, he was able to indulge in investigations inspired by pure scientific curiosity; but, on the other, the background of this curiosity was nevertheless determined by the interests of the people in whose environment he was brought up. He was led to study those aspects of the phenomena which might throw light on the possibilities of transformation of various natural agencies into each other and into mechanical power.

'Joule is not, of course, the only example of the influence exerted by the new industrial class. At about the same time as he engaged in his researches, Grove, a distinguished amateur of science, gave his famous lecture "On the correlation of the forces of Nature", at the London Institution. In this lecture the various aspects of the transformation of natural agencies into each other are treated in length, though only in a qualitative manner, and the novelty of this way of looking at natural phenomena is emphasized. It is significant that Grove's audience was not the Royal Society but the members of an Institution newly founded by the industrialists for the purpose of discussing scientific problems of special interest to them. It took quite

as to convince the learned world of it. Heat began to become a respectable part of physics. In 1859, Rankine, a great engineering teacher with a scientific bent, linked the idea of conservation of energy with a molecular theory of heat as a mode of motion and thus showed it need not be in conflict with Newtonian physics.

a long time before the new way of looking at Nature advocated by Grove and Joule was incorporated into the scientific tradition. This was not achieved before the middle of the century. There is a lapse of about eight years between the time when Joule formulated his main conclusion regarding the conservation of energy and his election in 1850 as a Fellow of the Royal Society, which indicates the complete change of outlook of the leading scientific circles.'

It is in the light of this that we must consider Professor Rosenfeld's earlier statement.

' . . . It would be quite wrong to say that Joule started his investigation of the connection between heat and mechanical power with a view to perfecting the engines providing such power. That could have been so, but as a historical fact Joule was primarily animated by pure scientific curiosity. . . . '

To be interested in the problems of utilization of natural forces is, as a social fact, deemed to be pure or not according to whether the person interested has not, or has the idea of making money out of the result. This may well turn not on the intellectual or moral quality of the person, but only on his circumstances. If he is above want or ambition of fortune then he can afford to indulge pure scientific curiosity. If he has to make money for himself or is able to do his work only as the employee or client of one who needs it, then his scientific curiosity can no longer be disinterested. Nevertheless, from the point of view of the history of science, the personal motive is less important than the fact that a phenomenon is investigated as if it had a possible utility. That is how the progress of science in relation to technology appears to proceed, largely independently of the private morals and interests of those that advance it and is on the other hand extremely sensitive to positive and negative social pressures.

Heat and Energy

Thermodynamics. Clausius, Thomson, Helmholtz

Clausius generalized Carnot's work in the form of differential equations and laid the foundation of formal thermodynamics. The final acceptance of the new views was due to the work of two academic scientists, William Thomson and Hermann Ludwig Ferdinand Helmholtz, who fitted them into the accepted scheme of classical physics and almost succeeded in obscuring the fact that a great revolution in human thought had been achieved.

What Thomson did essentially in his paper 'On the Dynamical Equivalent of Heat' (1851, Roy. Soc., Edinburgh) was to combine the theories of Carnot on reversible cycles with that of Joule on the mechanical equivalent of heat. This had already largely been done by Rankine and Clausius. The effect of his presentation, however, coming from the Professor of Natural Philosophy of Glasgow[1] and Cambridge 2nd Wrangler, was to set the seal of respectability on the first and second laws of thermodynamics. It also introduced as a definite physical concept the term *energy* as equivalent to heat and work and to the older *vis viva* of Leibniz and thus gave to the first law the classical form of the Conservation of Energy.

In Germany, Helmholtz,[2] a young army doctor who had

[1] He had obtained the chair in 1846 at the age of 22 as a result of a vigorous campaign by his father, James Thomson, the Professor of Mathematics. See S. P. Thomson, *The Life of William Thomson, Baron Kelvin of Largs*, London, 1910.

[2] It should be made clear that Helmholtz, the son of a highly cultured but poor schoolmaster at Potsdam, became an army surgeon because this was almost the only means at that time of acquiring a scientific education. He belonged to the brilliant school that gathered round the great physiological teacher, Johannes Müller, one of the

yet to make his name as a physiologist, had already had the same effect. In his classical paper of 1847 'On the Conservation of Force'[1] he discussed all the various aspects of energy, mostly already covered in Mayer's papers. The one new idea he introduced was that in an assembly of bodies, interacting by central forces only, the sum of kinetic and potential energy, or of force and tension as he called them, remained constant. This, though really implicit in eighteenth-century dynamics, had the great merit of reconciling the new doctrine with the Newtonian tradition and thus ensuring its incorporation in academic science.[2] By the turn of the century, therefore,

first to emancipate himself from the mystical influence of *Naturphilosophie*. A group of these students, including Clausius, Kirchoff, Wiedemann, Tyndall, and Werner Siemens, joined with Helmholtz in founding the Physical Society of Berlin, which was to form the spearhead of the new movement of experimental science. Helmholtz himself was led to the consideration of the conservation of energy by his study of animal heat, which he measured in contracting muscles by the use of a thermocouple, a method that was still being used in a refined form by Professor Hill a hundred years later. It is interesting to note that Helmholtz's paper 'On the Conservation of Force' in 1847 was at first refused publication by Poggendorf in his *Annalen*, just as he had refused Mayer's five years before, and it had to be published privately.

Helmholtz's later work was largely on the physiology of the senses, he was the inventor of the ophthalmoscope, and his work on sound formed the scientific basis for the telephone, the gramophone, and the sound track. He became the leading physicist of the new Germany and was the first director of the Charlottenburg Technische Hochschule, founded by his life-long friend, Werner von Siemens.

[1] *Poggendorf's Annalen*, translation in *Taylor's Scientific Memoirs*, 1853, p. 114.

[2] The battle for the priority of this greatest physical discovery of

the logic of the steam engine had been recognized by the scientific world in the form of new universal laws fit to take their place beside those of Newton. A new commo-

the nineteenth century was a fierce one and was conducted with little regard to scientific manners. In Britain Joule and Thomson together with Helmholtz took most of the credit until in 1863 Tyndall in his Royal Institution lectures had urged the claims of priority of Mayer. Tait and Thomson took the unusual course of attacking his claim in the popular journal *Good Works*. Tyndall reasserted it, which drew from Tait a stinging attack in the *Philosophical Magazine* of 1863, p. 263: 'Does Prof. Tyndall know that Mayer's paper has *no claims to novelty or correctness at all*, saving this, that by a lucky chance he got an approximation to a true result from *an utterly false analogy*; and that even on this point he had been anticipated by Séguin, who, three years before the appearance of Mayer's paper, had obtained and published the same numerical result from the same hypothesis? Prof. Tyndall has quoted, without comment, our note on the subject. Does he recognize the truth of that note? If he does not, let him expose its errors; and we shall be happy to acknowledge our mistake, and to hail the additions to scientific knowledge which (involving at least a reconstruction, if not a destruction, of thermo-dynamics) must result from Mayer's statement if it can be shown to be true.'

Tyndall turns the tables on him by quoting from Helmholtz himself, '"The first man who correctly perceived and rightly enunciated the general law of nature (das allgemeine Naturgesetz) which we are here considering, was a German physician, J. R. Mayer of Heilbronn, in the year 1842. A little later (in 1843) a Dane named Colding presented a memoir to the Academy of Copenhagen, in which the same law was enunciated. . . . In England Joule began about the same time to make experiments on the same subject." You say that you "never intended to hint that Prof. Tyndall could have meant to put Mayer forward as having any claims to this great generalization". Bad as I am, you would not think of charging me with the grotesque folly. And still for nine years Helmholtz, who was a master in this field before you ever entered it, has stood self-convicted of the very absurdity. In the subsequent portion of the article referred to, you come to what

dity, *work*, had been created by practice and a new concept, *energy*, accepted in theory.[1]

Yet although the laws of thermodynamics arose from

you denominate "the grandest question of all". You put it thus:—"Whence do we immediately derive all those stores of potential energy which we employ as fuel or food? What produces the potential energy of a loaf or a beefsteak? What supplies the coal and the water-power without which our factories must stop?" These "grandest questions of all" were asked and answered by Dr. Mayer seventeen years before you wrote your article; and yet you never mention his name. You proceed, "Whence does the sun produce the energy which he so continuously and liberally distributes?" You then consider the various hypotheses framed to account for the permanence of solar emission, and develop the meteoric theory of the sun's heat. Mayer did the same fourteen years before you wrote this article in *Good Works* and six years before you published your earliest paper on the subject in the Edinburgh Transactions. There is not an idea of any originality in the whole of that paper that is not to be found in the memoirs of Mayer; and yet you do not give him an iota of credit in this article of yours in *Good Works*, the accuracy of which you have so trumpeted forth.'

This was too much for Thomson and produced a reply which seems the Victorian substitute for a challenge—

'To the Editors of the *Phil. Mag. & Journal*,

'The tone adopted by Dr. Tyndall in addressing myself is of a character, I believe, unprecedented in scientific discussion. It is such that I decline to take part personally in any controversy with him, or to notice his letter further than to say that on all the scientific questions touched upon in it I am ready to support the correctness of the opinions, and information, in the article in *Good Works* by Prof. Tait and myself, and that I shall do so when I see proper occasion,

I remain, Gentlemen, Your obedient Servant,
William Thomson.'

[1] See p. 42.

consideration of the genesis of mechanical work—animal, chemical, electrical, and most of all from heat—it cannot be claimed that their formulation led to any immediate change in the practice of power production. Steam engines continued to be developed along lines indicating engineering improvements rather than any logical thinking out of the application of the new principles. The exploitation of electrical energy had to wait another thirty years. This is not very surprising considering the fact that engineering schools were in their infancy and that most engineers of the time were self-taught men who found enough scope for their activities in the rapid expansion of industry, where profits were so easily come by that efficiency was at a discount. The one significant scientific attempt at a Carnot cycle engine, Sir W. Siemens' superheated steam regenerative engine of 1858, was not a success. Where such a financial and technical genius failed others were not likely to succeed. Even to this day there is only one heat engine, the Philip's air engine of 1942, that has been inspired by purely thermodynamic conceptions, and even that has not come into general use.

The practical value of thermodynamics has been rather as a general guide to the design of the new internal combustion and turbine engines, whose basic mode of working was determined by mechanical possibilities. Where thermodynamics has been used most directly and successfully is in modern chemical engineering and in the reverse of the heat engine—refrigerating machinery[1] and heat

[1] The history of refrigeration deserves a chapter to itself because of the multiple links it shows between science, techniques and human needs.

pumps—of which only the former came into use before the end of the century.

On the other hand the introduction of the concepts of energy and the associated *entropy* of Clausius was to alter, once and for all, the whole mental outlook not only of physical scientists but also of biologists. They were to prove to be great unifying principles. In the hands of Gibbs and le Chatelier they were to be applied to the chemistry of inorganic and living systems, in those of Thomson, Kirchoff and Boltzmann to the understanding of electricity, light and all the newer forms of radiation.

Over and above the practical applications and theoretical developments of the first and second laws of thermodynamics is their deep and pervasive influence on all thought throughout the last hundred years. This influence was on balance a liberating one. Even the limitations they imposed on available energy only served to bring out how much of what was available was wasted. It is true that in the latter half of the nineteenth century the second law of thermodynamics was used to give scientific justification to a conservative despair. The caution and

The use of cold for preserving food is at least as old as the Stone Age and it was maintained throughout antiquity and the Middle Ages wherever snow and ice could be found or kept. Nevertheless, in most of the civilized world it remained a luxury. It was only when the wealth accumulated in great cities made it pay to attempt to produce it on a large scale that science was called in. Later still it was to be a major means of providing great industrial populations with food from far-distant colonies. On the one hand its development links with thermodynamics and the liquefaction of gases from which so much of modern physics flows, on the other with the study of fermentation and putrefaction and the rise of biochemistry.

pessimism of Lord Kelvin, as Thomson had by then become, led him into using it to prove that the sun must be cooling so rapidly that there could not have been enough time for evolution and that the universe was in danger of a general heat death. The discovery of radioactivity has dissipated that prophecy, at least for the time being, and men are beginning to see that what limits their control of the universe are social rather than material factors.

The idea of energy has gradually come to take the place of that of matter as the source of human well-being or even existence. The wealth of countries is measured not by the gold they own but by the number of horse-power available to each of their citizens. The winning of material wealth itself, food from the fields, metals from the ground, is seen to depend on the energy available. Coal, oil, the sun and now the atom are valued for the energy they can produce. All this would be unthinkable but for the work of those who in the early nineteenth century separated out the concept of energy from the confusing variety of its manifestations. As principles the first and second laws of thermodynamics are unshaken and have only been deepened by the limitation of energy exchanges by Planck's quanta and extended by Einstein's derivation of the equivalence of mass and energy.

Chapter III

FERMENTS AND MICROBES

CHEMISTRY, far more than physics, was the dominant science of the nineteenth century. This is so, in spite of the fact that the major physical discoveries found their development and application in the steam engine at the beginning, and electric power at the end, of the century. With chemistry, however, there was a far larger number of new processes that could be turned more immediately to profitable use, and this afforded directly and indirectly for the training and employment of an ever-increasing number of chemists. Indeed from the beginning of the century and increasingly till its end the chemists were the most numerous of the newly differentiated groups of scientists. Here Scotland was in the van. A Chemical Society,[1] the first in the world,

[1] The advent of separate scientific societies was itself symptomatic of the transition from the diffusely interested amateurs, the curiosi or virtuosi of the seventeenth century who founded the Royal Society, to the specialist professional scientists characteristic of the twentieth century. The transition was already foreshadowed in the active circles of science in the eighteenth century and its justification had already been given by Priestley in 1767.

'At present there are, in different countries in Europe, large incorporated societies, with funds for promoting philosophical knowledge

was in existence in Edinburgh before 1785. In 1814 Sir John Sinclair could write 'At present there are a greater number of intelligent practical chemists in Scotland in proportion to the population than perhaps in any other country of the world.'[1] The absolute number was however greatest in France, though England was not far behind. The Chemical Society of London was founded

in general. Let philosophers now begin to subdivide themselves, and enter into finaller combinations. Let the several companies make small funds, and appoint a director of experiments. Let every member have a right to appoint the trial of experiments in some proportion to the sum he subscribes, and let a periodical account be published of the result of them all, successful or unsuccessful. In this manner, the powers of all the members would be united, and increased. Nothing would be left untried which could be compassed at a moderate expense, and it being *one person's business* to attend to these experiments, they would be made, and reported without loss of time. Moreover, as all incorporations in these smaller societies should be avoided, they would be encouraged only in proportion as they were found to be useful; and success in smaller things would excite them to attempt greater.

'I by no means disapprove of large, general and incorporated societies. They have their peculiar uses too; but we see by experience, that they are apt to grow too large, and their forms are slow for the dispatch of the *minutiae* of business, in the present multifarious state of philosophy. Let recourse be had to rich incorporated societies, to defray the expense of experiments, to which the funds of smaller societies shall be unequal. Let their transactions contain a summary of the more important discoveries, collected from the smaller periodical publications. Let them, by rewards, and other methods, encourage those who distinguish themselves in the inferior societies; and thus give a general attention to the whole business of philosophy.'

Joseph Priestley, *The History and Present State of Electricity*, London, 1767, Preface.

[1] Clow, *op. cit.*, p. 599.

71

in 1845 with 77 fellows; in 1900 it had 2292, and now more than half the scientists in the country are chemists.

The major chemical discoveries of the role of oxygen in combustion and of the conservation of matter, and their explanation in terms of the atomic theory, belong historically to the end of the eighteenth century. They furnished at the same time the key to the unravelling of the constitution of natural and artificial bodies and an excitement and interest in its use. This attracted to chemistry, particularly in France under the combined influence of Lavoisier and the Revolution, the most brilliant minds of the new century. By the middle of the century the centre of interest, for reasons we shall discuss, had shifted to Germany where it was to remain till 1918, but the part played by Britain, though very uneven, was not negligible.

Despite the enormous volume of work done and the considerable industrial, agricultural and medical results that followed, the period was marked by only one radically new idea of scope or importance comparable to that of the conservation of energy in physics, or of evolution in biology. However, that grand conception—the periodic table of the elements of Mendeleev—contributed little to chemistry in its time, as its *raison d'être* could only be understood in terms of twentieth-century atomic theory. Nevertheless, the century did witness an immense progress and triumph both in technique and logic, so that from the few dozen simple compounds whose atomic constitution could be roughly guessed at the beginning of the century it passed by the end to the determination of precise structural formulae of some 20,000 complex compounds. The task of the century was to use the methods handed down by Lavoisier and Dalton to analyse and synthesize step by step every type of compound found in nature and to create almost as many new ones. In the process new methods and new ideas were added from time to time and incorporated to form a practical and intellectual weapon for the solution of new problems.

As we have seen, the chemical industry, though born from a union between the physician's and the miner's art, was all through the nineteenth century largely an ancillary to the rapidly expanding textile industry. Second only in importance was the impetus provided by the traditional industries of food and drink, pickling, baking, brewing and distilling. These remained, however, outside the orbit of science until they changed from a household or village basis to the scale necessary to satisfy the needs of the new industrial towns. It was then that new problems arose, which tradition found it hard to solve. Though at first it was science that had to learn from practice it soon returned the gift. The first contact was in physics rather than chemistry, for the chemistry of the time and of long after, could give little help—as the sequel will show.

The behaviour of whisky stills[1] had given Black the idea

[1] Strictly speaking, distilling is the first really scientific industry though its genesis lies outside the scope of this essay. The still was really a laboratory apparatus—the ambix or alembic—afterwards blown up to industrial dimensions. The Arabs used it for distilling rose-water, the Christians for aqua vitae, usque bagh or fire-water. The development of this still, which was very rapid from the mid-fourteenth century, when the drinking of spirits became popular, to the mid-sixteenth century coincided with the first rise of experimental

of the latent heat of vapours and through him Watt hit on the separate condenser which revolutionized the steam engine. It was Mr. Thrale, Samuel Johnson's friend, who first used a thermometer in brewing. Later, when chemistry had advanced far enough to be able to give some practical aid, chemists were called in to cope with the all-too-frequent cases where the change to large-scale and rapid operation led to unpleasant and unprofitable results.

At the risk of oversimplifying a story that would require several volumes to expound, I will attempt to bring out some of the principal steps that led from the elementary chemistry of the beginning of the century to the bacteriology and biochemistry of its end. To do this I have chosen two of the principal figures of nineteenth-century chemistry around which the most decisive controversies on the relation between the animate and the inanimate were centred. Justus von Liebig (1803–73) and Louis Pasteur (1822–95) were both men of their time. Though often antagonists they were alike in their enormous vitality, wide interests and, perhaps most of all, in their concern and passion to see their science applied for human welfare.

Liebig

Liebig, the son of a druggist at Darmstadt, is important less for his contributions to chemistry, great as they were,

chemistry. From then on until the Industrial Revolution it became stereotyped—a routine operation beneath the notice of the man of science. See Clow, *op. cit.*, and R. J. Forbes, *Short History of the Art of Distillation*, Leyden, 1948.

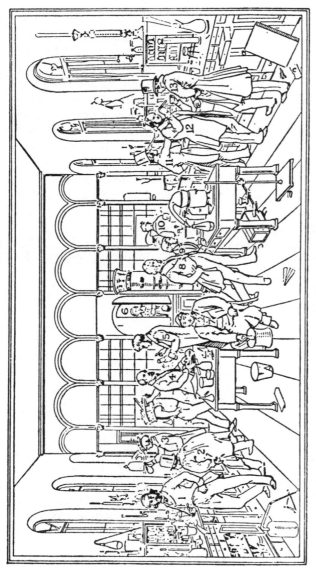

FIG. 3. Liebig's famous laboratory at Giessen. Among the numbered characters are many of the most famous chemists of the nineteenth century. Number 13 is A. W. von Hofmann. From John Read's *Humour and Humanism in Chemistry*, by courtesy of G. Bell and Sons, Ltd.

than for being the first to organize the teaching of practical chemical research and of popular chemistry, and because he preached and practised the application of chemistry to industry and agriculture. In his youth he had found no means of satisfying his thirst for chemical knowledge in Germany and was drawn to Paris, which in 1822 was the centre of all chemical advance. There he rapidly learned the latest methods and ideas. In two years, at the age of twenty-one, his work was so outstanding as to gain for him the chair of chemistry at the small university of Giessen. He taught there for twenty-eight years and in that period he made of Giessen the first chemical school, not only of Germany, but of the world, one on which all modern teaching and chemical laboratories are based. But he did not limit his teaching to the university. By his popular lectures and books, by his practical attention to agriculture and food chemistry, Liebig was to lay the foundation of Germany's chemical pre-eminence. With the new and accurate analytic methods which he developed himself and through the work of the brilliant students whom he attracted round him, Liebig attained in the forties a dominant position in the chemical world. He was early linked in co-operation with Wöhler, trained in the Swedish school of Berzelius, the only rival to Paris in the chemical world.

By the end of the thirties leading chemists had come to recognize that there was nothing mysterious or fundamentally different about the chemistry of living matter. This is traditionally associated with the preparation in 1828 by Wöhler of urea, a typically organic substance, from ammonium cyanate, which was taken as inorganic, though ultimately just as much derived from organic sources. Actually the idea was the result of many different experiments and represented the turning away of the new generation of chemists under French influence from the prevalent early nineteenth-century German *Naturphilosophie* with its essentially medieval inspiration of the kingdoms of nature. The chemical identity of the substances and laws of the organic and the inorganic were to have a profound philosophic bearing.[1]

What Liebig and his students set out to do was to explore the whole range of natural products, to analyse them and to try to reduce them to some kind of order. This led him increasingly to deal with the practical problems of agriculture, food and drink. With almost missionary zeal, he analysed and experimented and preached to farmers and manufacturers on the application of chemistry to their operations.

Agricultural chemistry

He had been induced to make a systematic attack on agricultural chemistry by a request of the British Association in 1839 to prepare a report on 'Chemistry and its Applications to Agriculture and Physiology', which he

[1] As Engels put it in the *Dialectics of Nature*, pp. 11–12:
'The wonderfully rapid development of chemistry, since Lavoisier, and especially since Dalton, attacked the old ideas of nature from another aspect. The preparation of inorganic means of compounds that hitherto had been produced only in the living organism proved that the laws of chemistry have the same validity for organic as for inorganic bodies, and to a large extent bridged the gulf between inorganic and organic nature, a gulf that even Kant regarded as for ever impassable.'

completed in the following year. This request is in itself symptomatic of the close relations between that body and German science. Actually, as we shall see later (pp. 139 ff.), the movement that gave rise to the British Association was, itself, the result of the awakening liberal movement in Germany. Its parent body was the Deutscher Naturforscher Versammlung, inspired by Oken, which first met in Leipzig in 1822. If it had done nothing more, it would have justified itself by the result of its initiative with Liebig.[1]

Basing himself on the work of the great pioneers, Priestley, Lavoisier and Davy, Liebig further elucidated the main chemical cycles of living matter. He showed that plants did not get their nutriment from a somewhat mystically conceived 'humus', however important it might be otherwise for the fertility of the soil, but from the carbon dioxide of the air and from the metals and salts dissolved in the soil, which were returned to it in manure. Here, however, he missed an essential point assuming that the nitrogenous component was also drawn from the air, consequently failing to see the basic importance of nitrates and ammonia. It is therefore not surprising that his commercial venture, the Patent Mineral Manure Company, though it was run by James Muspratt the pioneer of British chemical manufacture, was a failure.[2]

---

[1] Marx at least thought so. Writing to Engels in 1866 he complains: 'I had to wade through the new agricultural chemistry in Germany, especially Liebig and Schonbein, who are more important in this matter than all the economists put together'. (*Marx and Engels: Selected Correspondence*, London, 1943, p. 204.)

[2] See Clow, *op. cit.*

The constituents of plant ashes, containing no nitrogen and with ill-balanced salts, were expensive to compound and were only suited for a very limited class of soils. Nevertheless the attempt was in the end not an unfruitful one, for it was Liebig that inspired the wealthy landowner J. B. Lawes to start field manure experiments in 1834 on his estate at Rothamsted, the first British experimental agricultural station. Later, associated with Gilbert who had studied under Liebig, he manufactured superphosphate, the first artificial fertilizer, which was manufactured commercially by Richardson in 1844. By the end of the century the production of artificial fertilizers was to become one of the largest branches of the chemical industry and already superphosphate, kainite, and basic slag had become almost essential to heavy crop production, while ammonium sulphate was beginning to oust the natural Chile saltpetre as a source of nitrogen.

Liebig's influence on animal and human nutrition was equally far-reaching. He divided the nourishment of animals and man into the now classical—fats, carbohydrates and proteins—and brought out how the first two were burned up as fuel while the last provided for the repair of the mechanism. This proved a sound quantitative basis for the whole of agricultural and nutritional chemistry. It was not, as we know now in the age of vitamins, the whole story, but it was an enormous step forward.

Fermentation

Liebig's strength—his insistence on quantitative analysis and his belief in the dominance of chemical law—concealed a weakness, it had no place for living processes as such. With Wöhler he believed he had laid the spectre of a *vital* force which had always been invoked to *explain*—and thus prevent examination of—the *spontaneous* chemical changes of growth and decay. It had never been invoked more often than in the processes of fermentation where the 'archaeus' of Paracelsus had only changed the name and made precise the functions of the 'entelechy' of his hated Aristotle.

Liebig devoted much attention to fermentation, not only because of its growing industrial importance, but because it was still the only way of preparing so many organic compounds. He gave it a purely chemical explanation. For him it was primarily an oxidation—all bodies had to be exposed to the air before they fermented. Even alcoholic fermentation, which involved no addition of oxygen, took place only under the influence of partially oxidized *putrescible* material—the necessary yeast—which passed on energy by contact to *fermentiscible* material—sugar. Any organisms which might appear, such as the torulae which Caignard de la Tour had described in 1839, were purely coincidental or merely furnished putrescible material.

Pasteur

Here Liebig was challenged by a younger man, Louis Pasteur, who, with the same passion for science and its application, took the radically different view that the living yeast was the essential agent of the fermentation process. This was not because he was himself a biologist. Far from it, he started his scientific career as physicist and crystallographer, though he was also well grounded in chemistry by his teachers at the École Normale.

Pasteur's scientific career contains example after example of the interaction of scientific discovery and technical progress. On graduating, the first problem he chose for himself was the solution of the apparent contradiction that the two different acids, tartaric acid and racemic acid, should give salts of apparently identical crystal form. His triumphant separation, in 1848, of sodium potassium racemate into right- and left-handed forms set him off on a scientific career that was to lead him into fields widely remote from this apparently academic problem. As a matter of fact the whole of the investigation of racemic acid arose out of an accidental by-product of the tartaric acid industry, itself a commercial application of Scheele's preparation of the vegetable acids fifty years before.[1]

Pasteur's key discovery proved that molecules had definite shapes representable in space and thus laid the basis for stereochemistry. However, owing to intrinsic confusions about atomic weights and even more to a marked reluctance to admit the material existence of atoms, the first spatial representation of a molecule had to wait for Le Bel and Van't Hoff twenty-six years later. Pasteur followed another track. He had observed that a little mould, *penicillium glaucun*—a cousin of *penicillium notatum*, from which so much was to come in our time—attacked the right-handed and not the left-handed form of the acid. From this he deduced that life must operate asymmetrically, and that fermentation, which operates on the asymmetrical molecules of sugar, must be a *vital* process.

[1] A fuller account of this discovery and its antecedents and consequences is given in the second essay of this book.

The son of a tanner, coming from Arbois in the wine-growing district of the Jura, Pasteur had always been interested in fermentation. He received another impetus when, in 1856, as Professor of the new Faculty of Science at Lille, he threw himself with enthusiasm into the service of the chemical industries of the district—largely distilling and vinegar-making. It was a failure of a Mr. Bigo in making beetroot alcohol that led Pasteur to take the decisive step that drew him away from physics and chemistry into the unknown field of microbiology. He used the microscope to distinguish between the round yeast globules of alcoholic fermentation and the long vibrios of the unwanted lactic fermentation. He had at once found a practical test that any brewer could use and had started the study of the minute organisms responsible, he firmly believed, for the chemical processes of fermentation.

Back in Paris, in 1859, as Professor of the École Normale, Pasteur's work became for the moment more academic. He found himself involved in a most violent controversy on the nature and origin of fermentation, which he conducted with enormous gusto, and not merely with words but with experiments. If fermentation was due to organisms, how was it that it appeared *spontaneously* in fermentable materials? Because of germs from the air claimed Pasteur. Immemorial tradition, backed by the authority of the great Liebig and supported actively by a vigorous controversialist, Pouchet, supported the idea of fermentation as an inherent property and deduced the *spontaneous* origin of the organisms found in fermenting preparation. In proving they were wrong, Pasteur

devised most of the modern techniques of sterilization[1]
and asepsis and almost as soon turned them to use in sug-
gesting the method of 'Pasteurization' for saving the wines
of his native Arbois from spoiling.

The germ theory of disease

Already Pasteur was thinking of the possible relation of
germs to disease.[2] However, very characteristically for
the nineteenth century, it was not the diseases of man with
which he was to start—they were guarded from the impact

---

[1] Here, as so often happened in the nineteenth century and earlier,
academic discussion was only taking up a question that had been
solved in practice many years before. Vegetable and meat had been
preserved by boiling and storing in air-tight vessels by the chef
Appert, who published the method in his book *L'art de conserver pen-
dant plusieurs années toutes les substances animales et vegetales*, Paris,
1810; or if we are to believe Séguin, by Montgolfier, years before.
See also Clow, *op. cit.*, and A. Betting, *Appertizing*, San Francisco, 1937.
Soldiers, sailors and explorers were already living on canned food.
Nevertheless it would be very wrong to disparage Pasteur's achievement on this account. The experience of food preservation remained
within the ambit of the food industry. Pasteur generalized it and
applied it to the problems of wounds and diseases.

[2] In 1862 in a note to the Minister of Public Instruction (*Œuvres*,
Vol. 7, pp. 3–7) after talking of the universality of germs in the atmosphere he adds: 'How wide and useful to pursue is the field of these
studies, which bear in so close a relation to the different illnesses of
animals and plants, and which form certainly the first step in the
desirable path of serious research into putrid and contagious
diseases.'
This was not the first time this idea had been expressed. Boyle had
written nearly two hundred years earlier in his 'Essay on the Patho-
logical Part of Physik': 'And let me add that he that thoroughly
understands the nature of ferments and fermentations shall probably

of science by the redoubtable profession of medicine—
nor even those of farm animals, the preserve of the veteri-
narian. The beginning was to be made in a field which
offered the most immediate monetary reward for success
and was guarded by no profession. In 1865 an epidemic
among the silkworms threatened to extinguish one of the
major industries of France. All methods of dealing with
it had failed when Dumas, the high priest of French
science, sent Pasteur, though without funds or assistants,
to deal with the situation.[1] Knowing nothing of insects,
not even that silkworms turned into moths, Pasteur used
his eyes, his intelligence, and his experience of fermentation to such good purpose that in two years he had not
only found the cause of the disease but had evolved
practical methods for wiping it out. Out of all this, which
must have brought millions to the growers,[2] he got
nothing but polite and sincere thanks.

This story brings out a paradox in the history of science
which has been the source of much misunderstanding.

be much better able than he that ignores them, to give a fair account
of divers phenomena of several diseases (as well fevers as others),
which will perhaps be never properly understood without an insight into the doctrine of fermentations.' Tyndall later drew Pasteur's
attention to this passage.

[1] Pasteur recalls (*Œuvres*, Vol. 6, p. 447) that when he told him
he knew absolutely nothing of silkworms Dumas replied: 'So
much the better, you will not have any ideas other than those that
come from your own observations.' Pasteur comments on the right-
ness of this judgement: 'Yes, it may be useful to leave the marked
roads and hack out new paths, that is how one finds new horizons.
It is harder work but it carries a more personal and original
mark.'

[2] Pasteur estimates himself (*Œuvres*, Vol. 7, p. 20) that the annual

Pasteur himself was clearly ·financially disinterested throughout. When he asked for money he wanted it for research not for his personal use. Nevertheless, throughout most of his working life, as in this case, he worked on problems of immediate economic interest. Similar stories of the fight against the potato blight or the phyloxera of the vines, as told in E. C. Large's *Advance of the Fungi* (1940), show scientists brought in only when disaster threatens. The fact is that it is only where there is a powerful economic motive operating in an industrial or agricultural field that anyone thinks of a scientist or lets him have a look in. Here the scientist's motives are certainly mixed, but the mixture is not the crude one— science versus self-interests—imagined by some superficial critics of materialism; it is one of glory of science and benefit to society. Certainly Dumas, when he invited Pasteur to go to the silk districts, had both in mind. Actually such direct interventions of the scientist into economic matters were rare until the present century. The factors directing science along the same tracks as that of dominant economic developments were usually of a more general nature concerned with the flow of endowments, the opportunities for students and the general social atmosphere in a way which will be discussed further in the conclusion of this essay.

saving actually achieved by following his methods was of the order of 100,000,000 francs or £4,000,000, a substantial sum in those days. As to his own attitude he told Napoleon III, who asked him why he did not turn his discoveries to legitimate profit: 'In France scientists would consider they lowered themselves by doing so.' R. Vallery-Radot, *The Life of Pasteur*, London, 1920, p. 129.

Ferments and Microbes

The needs of science

Pasteur still had no proper laboratory of his own, but at last in 1868 he felt impelled to write an article for the *Moniteur*, then the official paper, making a plea for laboratories to save French science. The article was refused, but it was shown to the Emperor Napoleon III and published as a pamphlet. It shows how close the problems of science, industry, agriculture, and medicine came in Pasteur's mind. 'The boldest conceptions,' he wrote, 'the most legitimate speculations can be embodied but from the day when they are consecrated by observation and experiment. Laboratories and discoveries are correlative terms; if you suppress laboratories, physical science will become stricken with barrenness and death; it will become mere powerless information instead of a science of progress and futurity; give it back its laboratories, and life, fecundity and power will reappear. Away from their laboratories, physicists and chemists are but disarmed soldiers on a battlefield.

'The deduction from these principles is evident: if the conquests useful to humanity touch your heart—if you remain confounded before the marvels of electric telegraphy, of anaesthesia, of the daguerreotype and many other admirable discoveries—if you are jealous of the share your country may boast in these wonders—then, I implore you, take some interest in those sacred dwellings meaningly described as *laboratories*. Ask that they may be multiplied and completed. They are the temples of the future, of riches and of comfort. There humanity grows greater, better, stronger; there she can learn to read the works of Nature, works of progress and universal harmony, while humanity's own works are too often those of barbarism, of fanaticism and of destruction.

'Some nations have felt the wholesome breath of truth. Rich and large laboratories have been growing in Germany for the last thirty years, and many more are still being built; at Berlin and at Bonn two palaces, worth four million francs each, are being erected for chemical studies. St. Petersburg has spent three and a half million francs on a Physiological Institute; England, America, Austria, Bavaria have made most generous sacrifices. Italy too has made a start.

'And France?

'France has not yet begun. . . . Who will believe me when I affirm that the budget of Public Instruction provides not a penny towards the progress of physical science in laboratories, that it is through a tolerated administrative fiction that some scientists, considered as professors, are permitted to draw from the public treasury towards the expenses of their own work, some of the allowance made to them for teaching purposes.'[1]

---

[1] Vallery-Radot, *op. cit.*, p. 152. His own letter to the Emperor was less rhetorical but more to the point. In it we find (*Œuvres*, Vol. 7, pp. 10–11):

'If it is a matter of searching by a patient scientific study of putrefaction some principles that can guide us in the discovery of the causes of putrid and contagious diseases . . . I should need a spacious installation. How is it possible to research on gangrene, viruses, inoculation . . . without any proper place for receiving live or dead animals? Butcher's meat is extremely dear in Europe. It is in the way in Buenos Aires. How can one test in a tiny laboratory without resources, the methods which perhaps would make its preservation and

His wish was granted, but a few months later he had a stroke and, believing he would not recover, the authorities stopped the work. He did recover and began, after the disasters of the Franco-Prussian war, the great work on the conquest of disease that went on for another twenty-five years. Now he was no longer alone, he had brilliant pupils like Roux, Chamberland and Thuillier around him, and disciples abroad like Lister, as well as rivals, like Koch and his German school of bacteriologists, now all following the trail of germs. Even then the battle was not easy. Here again the economic advantages of curing animal diseases—anthrax, chicken cholera, swine fever—provided the next stepping-stones to the attack on the citadel of the medical profession. It had, however, at the same time the advantage of enabling Pasteur to discover by experiments the nature of the process of infection and immunization while accepting the failures, which with human patients might have stopped the work altogether.[1]

transport easy. Anthrax kills in Beauce annually stock worth 4 million francs. . . .

'It is high time to free experimental sciences from the poverty which is shackling them. . . . I propose to His Excellency the Minister of Public Instruction the foundation, under my direction, of a well-endowed laboratory of physiological chemistry.'

[1] It is a comforting thought to believe that this progress from animal to human diseases can be justified on humane grounds. In fact the opposition of the doctors to Pasteur, which often reduced him to impotent rage, held back prophylactic measures especially in hospitals for more than a decade. Lister, inspired by Pasteur, had begun his advocacy of antisepsis in 1865. Both of them, and a very few followers, realized that the appalling mortality from infected

88

Ferments and Microbes

In 1881, field tests showed the full success of the vaccine for anthrax in sheep. Now the attack on human epidemic disease could no longer be delayed. Pasteur and his school, in friendly competition with the German bacteriologists, proceeded to discover the organisms responsible for the dread diseases of cholera and typhoid, which had already taken two of Pasteur's daughters.[1] It

wounds and operations was preventable. Sedillot, one of Pasteur's few converts among the surgeons of the war of 1871, wrote 'The Surgeon's art, hesitating and disconcerted, pursues a doctrine whose rules seem to flee before research'.

When it came to vaccines for infective illnesses experiments on human subjects were absolutely necessary to establish their efficacy. The preparation of the way through animals was a move to defeat psychological and institutional obstruction rather than a scientific or humanitarian necessity. Vallery-Radot, *op. cit.*, p. 187.

[1] How deeply he felt on the subject was shown in his reply in 1883 to a certain Mr. Peter, an inveterate and persistant objector to the germ theory, who prided himself on treating diseases as entities and, in particular, on his opposition to the use of cold baths in treating typhoid fever.

'It is true that the moment when progress in medicine comes from an allied science, there start up these unconsciously reactionary minds who would go so far as to put their own special science under a protective ban. While affirming, as Mr. Peter does, aloud that they only wish to go forward they stiffen themselves against the movement which carries them away. . . .

'To hear him talk with such disdain of chemists and physiologists who deal with questions of disease you would think in truth that he speaks in the name of a science whose principles are founded on a rock. Do we need proofs of the slow advance of therapeutics? Here now for six months in this assembly of the greatest doctors, they are discussing whether it is better to treat typhoid fever by cold lotions than by quinine, by alcohol, or by salicylic acid, or even not to treat it at all. And when one is perhaps on the eve of resolving the question

89

appeared that all other infective diseases could be controlled by extension of the same methods. With his final successful attack on hydrophobia, which gave some of the first clues to the nature of virus diseases and immunity, Pasteur's life work was complete. He died in 1895, leaving for a new century the full harvest of his work.

Even with the story told so briefly, the interplay of scientific and economic interests at every turn is very evident. Both interests occur repeatedly, the discovery of some new scientific fact from the observation of an industrial process and the practical application of the results of experiment and theory. It should be apparent that both Liebig and Pasteur succeeded because both, in their different ways, felt strongly that it was not only necessary to advance knowledge but also to see that scientific advances were widely known and profitably used. They were both disinterested in the sense that they sought no personal profit, but not at all disinterested in securing the greatest social effect for their work. In this, both were anticipating the driving tendency of twentieth-century science. Both began to see their reward in their own time. Liebig witnessed the beginning of scientific agriculture with the use of fertilizer and of a rational large-scale food industry.[1] Pasteur had wider reward commensurate with the range of his interests.

of the etiology of this disease by a *microbe* Mr. Peter commits this medical blasphemy and says 'Oh, what do your microbes matter? It would only be just another microbe! . . .'

Œuvres, Vol. 6, p. 448.

[1] He is best known to posterity for the last and least of these achievements—'Liebig's extract of meat'.

Ferments and Microbes

The germ theory of fermentation and disease was already in the nineteenth century creating new possibilities of securing food supply and opening up new territories for exploitation. It gave a rational basis for food preservation, sanitation, and the control of epidemics. Quite as much as steel and electricity it was an essential element in the march of imperialism in peace and war. Without Pasteur the armies which took part in the two great tragedies of our age could never even have been assembled, much less maintained for years in the field.

The combined attack of Pasteur's bacteriology and Claude Bernard's chemical physiology was to transform medicine from a venerable tradition into a science. Now the doctor could, for the first time in history, intervene with understanding to help the curative processes of nature and sometimes succeed where nature alone was bound to fail.

The effect of Liebig and Pasteur on other sciences was hardly less great. Liebig, as a research worker and even more as a teacher, was one of the main founders of organic chemistry. Pasteur's first work on molecular symmetry provided the key to the spatial representation of atoms in combination, and is the basis of modern structural chemistry (see pp. 209 f.).

The great controversy of the living or non-living character of ferments has proved a most fruitful one. In the light of subsequent knowledge both the protagonists have been proved to have been right as well as wrong. Pasteur's microbes turned out to be the essential agents, not only for the fermentations but for Liebig's chemical cycle, the nitrogen cycle in the soil, where the humus was restored

91

to its important function. On the other hand Buchner showed, after Pasteur's death, that ferments exist as chemical entities (enzymes) inside cells[1] and that the living organism is not essential for their action but only for their formation. Modern biology is tending to rest more and more on the basis of chemical reactions directed by organically formed molecules, the joint work of Liebig and Pasteur.

[1] Pasteur came very near to resolving his apparent conflict with Liebig in his lifetime. He knew that soluble ferments existed. He was very familiar with the ferment of malt—diastase—which had been used by man for thousands of years. In 1875 he was able to confirm that the ammoniacal fermentation of urine was due to a soluble ferment but this he showed could only be produced by a living organism. He found no such soluble ferment given off by yeast in alcoholic fermentation and he just failed to see that the explanation was that in this case the soluble ferment was retained in the organism.

Chapter IV

THE AGE OF STEEL

IN the last third of the nineteenth century, economy and politics were largely influenced by the availability, cheap and in quantity, of what was virtually a new material—cast steel, which could be rolled or forged. Steel rapidly replaced iron in most of its important large-scale uses—for rails, for ships, for bridges and for buildings. Its superior strength and lightness not only made it cheaper for these purposes but also enabled much more daring constructions to be attempted. The adoption of steel was a major factor in opening the whole world to a new level of trade and exploitation in a matter of decades. It made colonial dependencies and undeveloped areas much more profitable, helped capital expansion, and was the material basis of the new imperialism of the end of the century. Economic and technical factors combined to make the steel age the prologue of a new period of wars and revolutions.

The transition from iron to steel appears on close examination to be due to the work of far fewer men than any other major transformation in techniques. Three great names stand out clearly, Bessemer of the converter, Siemens of the regenerative open hearth, and Gilchrist Thomas of the basic lining. The three changes they introduced, though extremely simple in underlying principle, marked each in its own way an abrupt and decisive change in technique. Yet they were more than technical changes, as witness the fact that despite the existence of a growing and enterprising iron and steel industry, not one of them came from anyone in the trade but instead from rank outsiders, whose ignorance of what could not be done was their chief asset. I have already commented[1] on the one significant addition to iron technology of the century, the hot blast, that was introduced by J. B. Neilson, a gas engineer. The technical break involved in large-scale steel production was also one too great to be bridged by slow improvements of existing techniques; it needed fresh and imaginative minds that did not know what could not be done.

Bessemer was an inventor and manufacturer of bronze powder; Siemens was an engineer whose chief interest had been in electricity and the steam engine; Thomas was a police court clerk with a classical education and a hobby of chemistry. What made their achievements possible was that they all, in very different degrees, made use of science. But this, in itself, is no explanation. There were, in their time, hundreds of engineers and chemists who knew far more science and even more of the science of metallurgy than they did. What they had in common, and what made their success, was their appreciation of the existence of technical and economic problems to which it was worth

[1] Footnote 1, p. 24.

applying science. They were determined to solve them by a radical examination of the ends to be achieved and not merely, as most metallurgists had done before them, by a mere modification of existing practice.

The approach of each of the three was very different, characteristic not only of the man but of his times. Bessemer, born in 1822, was an unusual combination of prolific inventor and self-confident enterprising capitalist of the mid-century. Sir W. Siemens, a trained engineer, was one of a band of brothers determined to apply science in one form or another to make their fortunes, characteristic of the new German capitalism of the seventies. Gilchrist Thomas, solitary and poor, was the only one of the three who appears to have been conscious of exactly what he was trying to do. He alone analysed the whole problem and found a theoretical solution which was successful in practice almost from the start. What is perhaps even more remarkable is that, unlike such inventors in the past, he had secured, in advance, a complete cover of patents, which even the most powerful interests were unable to break. In his work one can see the prototype of the industrial research establishment of the next century.

By 1854 the manufacture of iron, cast and wrought, was expanding with ever-increasing speed, but steel was still a luxury product, selling at about £50 a ton, of very uncertain quality and available in small pieces, as against £3–£4 for pig iron, and £8–£9 for wrought iron rails. It might have seemed that this disparity in price between products containing approximately the same materials would have led to an intense search for new methods of manufacture, but rightly or wrongly no ironmaster thought it was worth his while. The steelmakers on their side were as satisfied with a small turnover of a valuable commodity.

Bessemer and the converter

Henry Bessemer entered the field almost by accident, but was well equipped for the task. The son of an engineer and craftsman, he had little formal education but great practical experience, particularly of type casting and the reproduction of metals and works of art on which he supported himself in his early years. He had a genuine passion for mechanical work and had picked up a good deal of science, particularly chemistry. At the age of thirty he had had the idea of replacing the age-old hand process of making powder for gold paint out of leaf brass by one using machinery alone.[1] Realizing that everything

[1] What led him to do so is best told in his own words:

'My eldest sister was a very clever painter in water colours, and in her early life, in the little village of Charlton, she had ample opportunities of indulging her taste for flower-painting . . . and had, with much ingenuity, made a most tastefully-decorated portfolio for their reception. She wished to have the words—

STUDIES OF FLOWERS
FROM NATURE,
by
Miss Bessemer,

written in bold printing letters within a wreath of acorns and oak leaves which she had painted on the outside of the portfolio; as I was somewhat of an expert in writing ornamental characters, she asked me to do this for her, and handed me the portfolio to take home with me for that purpose.

'How trivial and how very unimportant this incident must appear to my readers. It was, nevertheless, fraught with the most momentous

depended on secrecy he devised an almost automatic plant
which he set up in a factory without windows, operated

consequences to me; in fact, it changed the whole current of my life,
and rendered possible that still greater change which the iron and
steel industry of the world has undergone, and with it the fortunes of
hundreds of persons who have been directly, or indirectly, affected
by it.

'The portfolio was so prettily finished that I did not like to write
the desired inscription in common ink; and as I had seen, on one
occasion, some gold powder used by japanners, it struck me that this
would be a very appropriate material for the lettering I had undertaken.

'How distinctly I remember going to the shop of a Mr. Clark, a
colourman in St. John Street, Clerkenwell, to purchase this "Gold
Powder". He showed me samples of two colours, which I approved.
The material was not called "gold" but "bronze" powder, and I ordered
an ounce of each shade of colour, for which I was to call on the following day. I did so, and was greatly astonished to find that I had to pay
seven shillings per ounce for it.

'On my way home, I could not help asking myself, over and over
again, "How can this simple metallic powder cost so much money?"
for there cannot be gold enough in it, even at that price, to give it
this beautiful rich colour. It is, probably, only a better sort of brass;
and for brass in almost any conceivable form, seven shillings per
ounce is a marvellous price.

'. . . Here was powdered brass selling retail at £5 12s. per pound,
while the raw material from which it was made cost probably no
more than sixpence. "It must, surely," I thought, "be made slowly and
laboriously, by some old-fashioned hand process; and if so, it offers
a splendid opportunity for any mechanic who can devise a machine
capable of producing it simply by power."

'I adopted this view of the case with that eagerness for novel inventions which my surroundings had so strongly favoured, and I
plunged headlong into this new and deeply-interesting subject.'

It was from old books, probably based on the *Treatise upon Divers
Arts* by Theophilus the monk in the eleventh century, or a derivative

only by his wife's three brothers. Thus he retained his secret for forty years. This achievement was in many ways more difficult than that of making steel and points to the factories of the age of mass production. As a result he made a small fortune, which he found essential for his large-scale experiments on making steel. He had, in fact, financed his own industrial research establishment.

Always combative, particularly when roused by official complacency and stupidity, he determined to attack the problem of the large-scale production of cast steel just because the War Office had maintained that it could not be done. He had devised an aerodynamically spun shell adapted to smooth-bore cannon which they said no cast iron gun could fire. He inquired why a steel gun could not be used and was told condescendingly that no such gun could be made, because uniform steel could not be had in quantity. He thereupon decided to produce it. His only qualification for the enterprise was his successful development, a few years before, of a method of continuous casting of glass sheet which is the basis of all present methods. This gave him a good knowledge of the making and management of furnaces. In his own words 'My knowledge of iron metallurgy was at that time very limited, and consisted only of such facts as an engineer must necessarily observe in the foundry or smith's shop; but this was in one sense an advantage to me, for I had nothing to unlearn. My mind was open and free to receive any new impressions, without having to struggle against

source that Bessemer found the principles of the process which was still carried out on a traditional basis in Germany. See *Sir Henry Bessemer, F.R.S., An Autobiography*, London, 1905.

the bias which a lifelong practice of routine operations cannot fail more or less to create.'[1]

He started with the idea of using the first scientific method of steelmaking, devised by Réaumur in 1720,[2] of fusing together wrought iron, almost carbon-free, and cast iron, saturated with carbon. This he proposed to do by using a reverberatory furnace, but had little success. Noticing that much of the gas remained unburnt, he introduced air pipes into the furnace itself. This increased the flame temperature but also, he noticed, burnt the carbon out of the iron. Seizing on this observation he was led, step by step, to blowing air through the molten metal in fixed and then in movable furnaces, and ultimately to his famous *converter*. This development, which took barely two years, was an admirable example of the combination of observation and full-scale experimentation with scientific explanation of each step. In 1856 he succeeded in turning out the first ton of cast steel from his converter and announced his success, characteristically enough for the times and the man, at the Cheltenham meeting of the British Association.[3]

[1] *Ibid.*, p. 136.

[2] Steel had been made by adding carbon to wrought iron by the blister steel or cementation process since the days of the Chalybes in the tenth century B.C., or even earlier. It had remained a deep secret till Réaumur set himself to find out scientifically how it was done. He succeeded and published the result in *L'Art de Convertir le Fer Forgé en Acier* with a fine preface scorning personal or national secrecy of technical processes. See J. D. Bernal, *The Social Function of Science*, 1939, pp. 150–2.

[3] It was not, however, thought to be of sufficient importance to be printed. There is a story told by Nasmyth that Bessemer overheard

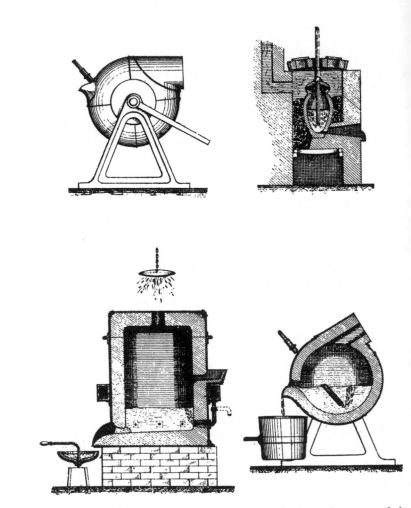

FIG. 4. Some of Bessemer's early converters. The iron plate suspended above the blast to prevent the fireworks going too far was completely melted in one of the trials.

Bessemer's process was enthusiastically taken up by the ironmasters, but almost as quickly dropped because of the apparently fatal flaw of failing to remove the phosphorus which most English and European ores contained. He himself, however, persisted and set up a works in Sheffield, and in the teeth of the steelmakers made steel with phosphorus-free ores from Sweden, Spain or Cumberland which was better than most and cheaper than any that they made.[1] The rest of Bessemer's business life was spent in fighting a long, but ultimately successful, battle against the prejudices and interests opposed to the use of the new steel. This, though it involved little science, was an essential part in the change-over from iron to steel. In the hands of a less combative and business-like man, the innovation might have been overborne and the transformation postponed for decades.

This surmise requires the usual qualification that inventions rarely come singly. About the same time a number of men were working towards the idea of violent oxidation of molten cast iron. Nasmyth had already in 1854 blown steam through molten iron to keep it stirred; he was within an ace of discovering the converter but he generously admitted Bessemer's claim to the full credit because he under-

an ironmaster say at breakfast in his hotel 'Do you know there is someone come down from London to read us a paper *on making steel from cast iron without fuel?* Did you ever hear of such nonsense?' (*James Nasmyth, Engineer: An Autobiography*, London, 1883, pp. 367–8.)

[1] Bessemer's quotation for high-class tool steel was £42 per ton as against £50 to £60 by other makers.

stood what he was doing.[1] William Kelly, a Pittsburg man, came closer still and made what would be now known as a side-blown converter in his Suwanee Furnace on the Cumberland river in which he made plates for the Ohio and Mississippi steamboats as far back as 1851. Lacking the knowledge, the facilities and the money that Bessemer could command, and caught by the instabilities of American finance, he was unable to develop his method and his patent was invalidated.

The practical success of the converter was assured once it became possible to control accurately the low but vital carbon content of the steel it poured. One way to achieve this, but a difficult and unreliable one, was to stop the blast at the right moment. A far better way was to burn all the carbon out and then to restore just the right amount at the end of the process. Spiegeleisen, an iron manganese carbide, proved to be an ideal way of doing this, as the manganese had an additional deoxidizing effect

[1] Bessemer's own account of Nasmyth's remarks seem worth quoting again:

'If I remember rightly, you held up a piece of my malleable iron, saying words to this effect: "Here is a true British nugget! Here is a new process that promises to put an end to all puddling; and I may mention that at this moment there are puddling furnaces in successful operation where my patent hollow steam Rabbler is at work, producing iron of superior quality by the introduction of jets of steam in the puddling process. I do not, however, lay any claim to this invention of Mr. Bessemer; but I may fairly be entitled to say that I have advanced along the road on which he has travelled so many miles, and has effected such unexpected results that I do not hesitate to say that I may go home from this meeting and tear up my patent, for my process of puddling is assuredly superseded."' *Ibid.*, p. 368.

and cured the defect of early Bessemer steel of being red-short. Its preparation and properties had been investigated by a skilled metallurgist, Robert Mushet, who proposed its use in steel making. He was the son of David Mushet, the discoverer of the black-band ironstone that was to make the fortune of the Scottish iron industry. Mushet's patent was badly drawn up and Bessemer was able to evade it, but he seems to have had some conscience in the subject for on being appealed to by Mushet's daughter he paid the now impoverished steelmaker a pension of £300 a year until his death in 1891.[1] There is no question, however, that it is to Bessemer and not to Kelly or Mushet that the real credit for making steel on a large scale is due. Yet, in spite of his early success, many obstacles still remained to be overcome before the full day of cheap steel.

Siemens. The regenerative open hearth

To many ironmasters the converter seemed an unnecessarily costly and complex machine and they welcomed a simpler way of making steel in large quantities. This was provided in 1867 by the Siemens-Martin process. It is difficult to place Sir William Siemens in the history of science and technology because, though by himself he would be a typical scientific inventor of the nineteenth century, he cannot be considered apart from his family of six distinguished and enterprising brothers, particularly the eldest brother, Werner von Siemens,[2] who saw to his education. It was the latter who directed what was vir-

[1] See F. M. Osborn, *The Story of the Mushets*, London, 1952.
[2] See p. 121.

tually an international syndicate of inventors and industrialists centred in Germany but operating in Britain, France, Russia and indeed all over the world. The Siemens family contributed perhaps more than any other group of persons to the creation of modern German industry and gave it, from the start, the academic scientific basis that neither British nor American industry acquired till very much later.

Nevertheless, the industrial level in Britain in the sixties and seventies was still higher than in Germany, and Sir William Siemens—the English Siemens—found there a wider market for the exploitation of the family's inventions. Though these were multiple and covered an enormous range, the most successful were the production of instruments and cable for the growing telegraph industry, the development of the dynamo and electric power production, and industrial furnaces. Unlike Bessemer, who attempted to solve a number of specific problems one after the other, using any means that occurred to him, what we now would call a series of *converging* researches, the Siemens were consciously applying certain general engineering principles and feeling their way as to where they could be used most profitably, typical *divergent* research.

The principle that concerns us here is that of *regeneration*. The idea is one of common sense backed by the thermodynamic principle that heat should not be allowed to escape from any industrial process but should be used for warming-up incoming products. Applied to metal furnaces it meant an end of the flaming chimneys that for nearly a century had lit up the sky in Sheffield and the

Black Country. Instead the hot gases heated a mass of bricks white-hot and escaped when relatively cool, while fresh air was later heated up by forcing it over the hot bricks. Originally intended as a means of economizing fuel, regenerative furnaces proved to have the unhoped-for power of raising temperatures to levels never before reached. Using gas fuel to avoid contamination they made the original Réaumur process of a mix of scrap iron and pig iron possible on a large scale and on an *open hearth*. Just that extra temperature was needed to convert the tedious puddling process, in which the purer, and consequently less fusible, iron was drawn out in a pasty condition, into one in which the whole charge could become a bath of melted steel which could be poured out by the ton.

There were, naturally, many practical difficulties, particularly in finding refractory furnace linings, but these were largely overcome by the French ironmaster, Etienne Martin. The final result by 1867 was a process as cheap and good as Bessemer's, but slower and far less mechanically and chemically revolutionary, thus appealing more to the traditionally-minded steelmakers. Siemens himself was not notably successful in his use of it. His own Landore works, though it turned out excellent steel, suffered seriously in the depression of the seventies and Sir William lost there much of what he had gained in his other enterprises.

Gilchrist Thomas. The basic lining

The regenerative open-hearth furnace did not depend as much as Bessemer's on phosphorus-free ore, but it still could not make steel out of the highly phosphatic ores of Cleveland and the by-then-German Lorraine,[1] which could only be used for iron. By the middle of the seventies this disparity was a glaring one and the problem of making steel from phosphatic ores was an open challenge. Yet despite all the new talent called into action in the steel industries of Britain, France, and Germany the solution did not come from any ironmaster or metallurgist, but as the result of the deliberate and carefully directed researches of a very part-time scientist, Sidney Gilchrist Thomas.

He came from a mixed Welsh and Scottish family with serious and intellectual tastes, particularly marked in his mother.[2] His father, a civil servant, was never very well-off and his early death left the family in very reduced circumstances. It was for that reason that Sidney, at seventeen, with a promising university career before him, felt he must give it up and obtain an extremely wearing job as a junior police court clerk in the London docks, the conditions of which probably conduced to his early death from tuberculosis. He continued his studies at Birkbeck College[3] in the evenings, concentrating on chemistry.

[1] The transfer of sovereignty was important because it enabled Lorraine ore to be worked with Ruhr coal without any of the difficulties of divided ownership and customs barriers.

[2] See L. G. T. Thompson, *Sidney Gilchrist Thomas*, London, 1943.

[3] Birkbeck College, set up in 1823 as the London Mechanics' Institution, was a typical product of the new independence and desire for learning of early nineteenth-century artisans. In that year some 2,000 mechanics and tradesmen met in the Crown and Anchor Tavern in the Strand and decided to subscribe to the hiring of rooms and professors. The principal object of the founders was to provide infor-

The Age of Steel

There he heard the lecturer, Dr. Chaloner, say that any-one who could solve the problem of making steel from phosphoric ores would make his fortune. Thomas needed that fortune badly, not, as might be thought, for the support of his mother and sisters, but because the horrors which he had to witness every day at his East End court filled him with idealistic schemes for social reform.

He, therefore, set himself deliberately to the problem, not as Bessemer had by observations based on large-scale trials, but by scientific analysis of the factors involved. He read voraciously in the technical literature and himself became a contributor to the journal *Iron*, using his scant leisure to visit ironworks at home and abroad. The solution of the problem which had baffled all professional

mation on 'the facts of chemistry and of mechanical philosophy and of the science of the creation and distribution of wealth'. It is not surprising that such a proposal, at the time when even the upper classes lacked acquaintance with these subjects, was treated as subversive if not revolutionary. The *St. James's Chronicle* wrote in 1825:

'A scheme more completely adapted for the destruction of this empire could not have been invented by the author of evil himself . . . every step which they take in setting up the labourers as a separate and independent class, is a step taken, and a long one too, to that fatal result.'

These alarms proved to be somewhat premature. The institute soon came under the control of liberal philanthropists like Brougham and Birkbeck himself, but it remained a place, and almost the only one, where science could be seriously studied by those in full-time employ-ment. Since 1920 a college of the University of London, it still pre-serves something of the character of its foundation in that the students elect two of the Governors—a privilege only secured after a vigorous political battle. See C. Deslisle Burns, *A Short History of Birkbeck College*, London, 1924.

metallurgists took him from 1871 to 1875, using a cellar as a laboratory and working in what time was left between his duties and his courses.

Essentially the solution was that of absorbing the oxidized phosphorus or phosphoric acid in a basic lining, made from magnesian limestone. The necessary practical tests were carried out in the course of 1877 by his cousin, Percy Gilchrist, who was employed as a chemist in the Blaenavon ironworks, working at first with a miniature converter holding eight pounds. Later with the assistance of the manager, Mr. Martin, they worked up to 11 cwt. By March of the next year Thomas was confident enough to take out patents and announce his discovery to the Iron and Steel Institute, where no notice whatever was taken of it. Again in September, at the Paris Exhibition of 1878, his paper on the new method was not found sufficiently interesting to be read, but there Thomas met Mr. Richards, the manager of a Cleveland ironworks, and impressed him enough to persuade him to make some full-scale tests. Finally on April 4th, 1879, the first few tons of steel were made at Middlesbrough.

Immediately the whole steel world was at his door clamouring for licences. Krupps tried to evade the patents, but Thomas, who had not been a police court clerk for nothing, brought them to law and forced them to pay up. The process, unlike Bessemer's, was an almost unqualified success from the start. It made the steel industry of Middlesbrough, of the Ruhr, and of the southern United States. The steel age had begun in earnest, but the man who had made that possible and who could have gone on to further triumphs had only a few years to live, dying in Paris in 1885 and leaving the bulk of a now considerable fortune to charity.

Science in the metal industry

In considering these three contributions to the nineteenth-century revolution in metallurgy, the first striking point is their independence of any organized scientific movement. Of the three inventors only Siemens had a university education, and none of them received any material assistance or more than a little advice from academic, scientific or governmental sources. This is the more surprising as by the mid-century the official scientific movement was in full swing, thanks in part to the German influence that came in with Prince Albert. In fact, the beginning of organized and subsidized industrial research had to wait another ten years, though its value had been amply proved by Bessemer's work.

In those days it was still thought sufficient to provide institutes and bodies like the British Association for promoting and discussing the reports of research carried out by private means. The inventor was expected to recoup himself by exploiting his invention under cover of the patent laws, if he was lucky. We see, in fact, Bessemer and Siemens using the results of one invention to finance the next, while Thomas was only able to work his up in the first place by pinching and starving himself. Even the military departments, which were to benefit so largely by the results of the revolution in steelmaking, did not encourage it and indeed were so attached to older materials and methods that they subjected the innovators, particularly Bessemer, to a series of painful snubs. These conditions were exaggerated in an industry like iron and steel, which had grown up in the hard way by expanding the operations of smiths, with the occasional injection of critical new ideas by outsiders, such as Roebuck and Cort in the eighteenth century, both of whom ruined themselves for their pains.

The result of this state of affairs was that new methods penetrated very slowly and there was an altogether unnecessary time lag between scientific knowledge and its application. None of the scientific ideas used by Bessemer, Siemens and Thomas were more recent than 1790 and the steel age might well have begun fifty years before it did. It would, however, have grown more slowly, owing to the far more limited technical and financial structure of capitalism early in the century. Equally well, but for the enterprise and good fortune of the inventors, particularly Bessemer, it might have been delayed, but not for so long, because it is clear from the methods of Siemens, and still more of Thomas, that science was soon to reach a stage when it could be used consciously and deliberately to solve industrial problems.

The Age of Steel

Steel came in at a time when capitalism was already fully developed. Indeed the processes involved in its making could hardly have been financed at an earlier period. Its advent was in turn to modify and transform still further capitalist production methods. The concentration already appearing by the mid-century in the iron industry was much enhanced by the rise of the heavy-steel industry. The capital required was far greater; only the largest firms could survive and even these were forced to amalgamate.

Steel kings, like Carnegie, Schneider, Krupp, and Vickers, began to dominate the industry and the stage was set for the cartels and monopolies of the next century. Already, well before the end of the century, steel in one form or another—rails, rolling stock, mining machinery —was making up much of the capital export that characterized the new *imperialism* by which the industrial countries were extending their grip on the undeveloped areas of the world.[1] The rivalries thus engendered, added

[1] J. A. Hobson in his classical book *Imperialism*, London, 1902, revised 1938, in the chapter on 'Economic Parasites of Imperialism', writes:

'What is the direct economic outcome of Imperialism? A great expenditure of public money upon ships, guns, military and naval equipment and stores, growing and productive of enormous profits when a war, or an alarm of war, occurs; . . .

'If the £60,000,000 (in 1905) [in 1938 it was £200,000,000, and in 1952 it was £1,634,000,000—J.D.B.] which may now be taken as a minimum expenditure on armaments in time of peace were subjected to a close analysis, most of it would be traced directly to the tills of certain big firms engaged in building warships and transport, equipping and coaling them, manufacturing guns, rifles, ammunition, 'planes and motor vehicles of every kind. . . . Here we have an important nucleus of commercial Imperialism. Some of these trades, especially the shipbuilding, boilermaking, and gun and ammunition making trades, are conducted by large firms with immense capital, whose heads are well aware of the uses of political influence for trade purposes. The public debts which ripen in our colonies, and in foreign countries that come under our protectorate or influence, are largely loaned in the shape of rails, engines, guns, and other materials of civilization made and sent out by British firms. The making of railways, canals, and other public works, the establish-

to the struggles for dominance inside Europe, led to a further demand for steel in the form of armaments. The malleability and toughness of the new steel compared with that of cast iron and the possibility of using it in large pieces enormously increased the size and effectiveness of weapons. Moreover, because it was needed both for ever-heavier guns and stronger armour to resist them, it also made armaments far more expensive and tended to make the state more and more dependent on a new heavy-metal industry. The age of steel was to prove far more warlike and disturbed than the first age of iron, so accursed by the poets of Greece.

ment of factories, the development of mines, the improvement of agriculture in new countries, stimulate a definite interest in important manufacturing industries which feeds a very firm imperialist faith in their owners' (pp. 48–49).

The tenacity with which capitalist interests in Britain cling to the control of the steel industry in an era of profitable rearmament, is shown by their insistence on reversing the very half-hearted Labour Party nationalization of the industry while making no attempt to do the same with coal, gas, electricity, or railways.

Chapter V

ELECTRIC LIGHT AND POWER

ELECTRICITY was not a new subject in the nineteenth century, but it was then that it was to receive its greatest theoretical extension and, in the form of the telegraph, to become an indispensable part of social life. Yet it was only at the end of the century, with the development of electric light and electric power, that its full practical capacities were beginning to be realized. The twentieth century was to be the electrical age. In electricity, unlike other aspects of technology, practice lagged far behind theory. The reasons for this were multiple, but as we shall see they were far more social and economic than technical.

The knowledge of electricity did not originate, as had that of chemistry, in an attempt to understand the uses men already made of natural forces. Natural electrical phenomena, though multiple, are not obviously connected and show themselves as trivial effects, like the attraction of amber for scraps of straw, or as the majestic but uncontrollable strokes of lightning. Indeed the study of electricity began only as a somewhat playful extension of that of its sister science, magnetism, which had already long proved its practical use in the compass. Gilbert was the first to recognize their relationship in 1600, but more than a century was to pass before a number of gifted amateurs took up the subject afresh and started a series of experiments that step by step elucidated its main principles. Among them was that great hero of the eighteenth century, Benjamin Franklin, who more than anyone else made electricity a science by linking the spark and the lightning flash and then fixing the knowledge in practice by inventing the lightning conductor.

A new era in electricity began in 1800, as already mentioned (p. 12), with the discovery of the electric current and of the voltaic cell needed to produce it. But at first only a fraction of the potentialities of the current—its heating and chemical effects—could be appreciated and the methods of generating it were limited. The first important step to break this limitation was made early in the century—the discovery of Oersted in 1820 of the effect of electricity on the magnet, which joined the two sciences of electricity and magnetism and gave them a common mathematical basis of theory. This link was completed by Faraday's discovery, eleven years later, of the complementary effect of the generation of electricity by a moving magnet. No new general principle entered into electromagnetism in the rest of the century. The main scientific effort throughout this period was devoted to the development of a quantitative and comprehensive theory, which culminated in Maxwell's great electromagnetic theory of light—the theoretical basis of the radio industry of the twentieth century. Some experimenters, it is true, studied the beautiful but complex phenomena of electric discharge and revealed, at the very end of the century, the existence of the electron, but its role, too, was to be in the future.[1]

In the nineteenth century itself, electricity was used for communication, light, and power, in that order. Everything that was done in these fields was implicit in the fundamental discoveries of Oersted and Faraday, but it took the whole of fifty years before these possibilities could be realized. Electric light came in during the eighties and the commerical use of electric power followed within a decade. The equipment needed for them did not require more engineering skill or capacity for research than was available in the thirties. Yet application, until the very end of the century, was halting and clumsy and remained on a very small scale. The delay can be put down to the combined effect of lack of vision and lack of funds for research and development. Brunel, one of the most far-sighted engineers of the mid-century, declared that electricity was only a toy. The few, like Faraday himself, who saw its possibilities lacked the inclination and the means to realize them. The actual advance came by a series of steps each of which had to pay its way before the next could be attempted. Most of the cost of electrical research had to be borne by the telegraph, the first commercially successful application of the new science.

The telegraph

The electromagnetic telegraph was only the latest of a series of abortive attempts to use other properties of electricity, such as the discharge of condensers and the

[1] Nevertheless, as will be shown, these researches were the means of improving the vacuum pump, without which the incandescent bulb would not have come into existence.

electrolysis of water, to send signals. The problem of telegraphy is really twofold; one aspect is the mathematical ingenuity in finding the most effective code for transmitting messages by unit signals, the other is the physical problem of sending and receiving those signals. The first was solved in 1832 by Morse who reduced the signal to its simplest element of dot and dash in a way so elegant that it still survives. The second was the fruit of the work of scores of electrical research workers, among whom are found the names of the most illustrious physicists, such as Henry, Gauss, Weber, and Wheatstone. The solution of the problems of current generation by batteries, of current propagation in networks of wire and of electrical measurement arose from and contributed to the development of telegraphy. It was in connection with these researches that both the methods and the standards of electrical measurement were evolved. This was to lead to the understanding of the fundamental relations between electricity and magnetism expressed in Maxwell's electromagnetical theory of light. It also led to the addition of the new electrical units to the age-old weights and measures for determining quantities of a new commodity that could be bought and sold.

The telegraph came in with the railway and was almost essential for its speedy working; it became a great personal convenience, but its chief influence was commercial. By linking markets together it made them into one vast market for commodities and stocks in which a change in price in one part affected the whole system at once. As the telegraph advanced it opened the world to capitalism. For long, however, it was limited to land and short sea routes; the markets of Europe and America were isolated from each other. The successful laying and operation of the transatlantic cable, which took the whole decade from 1857 to 1868, was a combined electrical, technical, and oceanographic triumph. To achieve it needed the full vigour, intelligence, and shrewdness of William Thomson—who rose to be Lord Kelvin on account of it. The details of the interaction of the scientific, technical, and financial problems it involved bring out most clearly the way in which science in the sixties was already being integrated into the economic system. Telegraph companies and later cable and telephone companies were the first purely scientific commercial enterprises. Nevertheless, though they gave employment to the new profession of electrical engineers[1] and provided some interesting problems for professors of physics, their direction was soon out of scientific hands and they contributed but a trifle of the vast profits they made to the advancement of science. The indirect material contribution of the telegraph to science was, however, an indispensable one. It provided the stock-in-trade for electrical experimentation—batteries, terminals, insulated wire (Faraday had to use wire from milliners or wind his own insulation), coils, switches, simple measuring instruments—and all at prices which even impoverished university laboratories could afford.

Though the telegraph was to be the nursery for elec-

[1] Edison claimed to be the first to have called himself an electrical engineer in 1870 but this is disputed. The Society of Telegraph Engineers in Britain was founded in 1871. Only in 1889 did it change its name to the Institute of Electrical Engineers.

trical technology the problems it posed were limited by its use of very weak currents. It needed delicacy and speed rather than power. The main line of advance was to come from the sale of a commodity needed even more than communication—domestic lighting.

The scientific problems of domestic lighting

Science had already owed much to the earlier chemical developments in this field. The discovery of the fatty acids and of the principle of atomic substitution arose from an increased demand for candles. Whale oil, and later paraffin, provided the means to a steady and brilliant flame. Both of these means of lighting were by mid-century relatively minor adjuncts of the omnipresent coal-gas industry, itself the basis of the new dyestuffs industry, and with it of organic chemistry.[1] Faraday's first paper was on the analysis of benzene found in gas mains.

The contribution of lighting to the electrical industry and to science was to be much more far-reaching. This was in the first place because, much as oil and gas were appreciated as compared to torches or candles, they seemed in a progressive age dim, smelly, and dangerous. For street lighting in particular something more brilliant was called for. There was, therefore, a premium on any alternatives and as the century drew to its close that premium increased with the phenomenal growth of cities.

The existence of this premium did not, however, lead straight away to any technical achievement. It merely assured that once a solution had been found it would be bound to be immensely profitable. Such hopes, however,

[1] See p. 14.

could not finance development; money for this had to be found, as will be shown, from the returns of limited solutions. There was neither the possibility nor even the idea of driving straight at the goal of the production of cheap, clean and brilliant light.

Nevertheless, blind and intermittent as was the advance to this goal, the technical and scientific knowledge gained on the way was of crucial importance, including as it did the generation and utilization of direct and alternating electric currents, together with the development of high frequency and vacuum techniques. It was to be the nursery of power and radio engineering, and to provide the experimental basis for modern physics.[1]

[1] Professor Polanyi has chosen the field of artificial lighting as an example of *applied science* to contrast most unfavourably with the science of mechanics as an example of *pure science* (*Logic of Liberty*, 1951, pp. 70–4). Leaving aside the omission of the fairly obvious contributions of practical men to the latter and the absurdity of comparing a vast field of pure with a corner of applied science the whole point of the comparison is missed. The satisfaction of a human need, whether consciously aimed at or not, has in the past been a focal point of many sciences, offering them inspiration and material support; it has not, and cannot be expected to have, of itself an inner logical coherence. In modern language the pursuit of lighting represents an area of *convergent* research in contrast to the *divergent* and multiple consequences of such a notion as statistical mechanics or such a material as tungsten carbide.

The deduction that Professor Polanyi is trying to draw is that such fields of applied research are inferior and not worth studying. Just because in the past, and particularly in the régime of free trade capitalism, whose passing Polanyi so regrets, such developments did show no rhyme or reason he presumes they could never do so. This is essentially obscurantist and ignores the whole method of directed research—convergent and divergent—of the present day. The under-

Electric Light and Power

The knowledge of the possibility of electric light, oddly enough, came before that of electrical communication or power. The first batteries, at the beginning of the century, were strong enough to make platinum wires white-hot— the prototype of the filament lamp—and to produce a persistent and brilliant light from two carbon rods which were touched together and separated—the prototype of the arc lamp. But batteries were too expensive to make and run for either of these forms of lighting to compete with the newly flourishing gas light.

The development of the dynamo

In principle this difficulty should have been removed by Faraday's magneto-electric generator, which could turn mechanical power into electricity. The development of this, however, hung fire in a way characteristic of competitive capitalism. The promoters of electric devices could not sell them because there was no cheap source of current. At the same time there was little urge to develop generators because there was no demand for the current. It was necessary to wait for demand and supply to come together before any rapid progress could be made. We can now see easily enough that the demonstration of magnetic generation of electricity brought it out of the region of merely transmitting information to that of transmitting mechanically useful power. At the time, however,

standing of human needs and the orderly and fully scientific search for them is just as proper a field of science as any other, only it is more difficult and less pleasing to vested interests. And in this there is still much to be learned from the past history of applied science, from its mistakes and failings as much as its successes (see pp. 147 f.).

only very few men, like Jacobi[1] and Joule and perhaps Faraday himself, could see that possibility, but as they lacked large resources, encouragement, and keen colla-borators, they were in no position to realize it. The tech-nical gap that had to be bridged was a considerable one—from apparatus operating on milliwatts to machines using kilowatts. This gap, however, could be bridged in stages. The first was the use of electric current in plating—a logi-cal application of Davy's work—put forward almost simultaneously by Elkington in England, Jacobi in Russia, and William Siemens in Germany, and later by Weston in America.

It was for electroplating that the first crude magneto-electric machines of Pixii were employed. The new stage was reached in the fifties when the arc lamp was first used in lighthouses, although the first permanent installation was the Dungeness light in 1862. The essential difficulty was that any machine depending on permanent magnets was bound to be inefficient for the heavy currents re-quired in arc lighting. From then on, however, a demand for a better machine was established.

The first important stage in the advance came from Werner von Siemens, working on very different appli-cations. Trained as a Prussian army officer, young Werner had been a pioneer in introducing the telegraph first into the army and then to civil use. He was also responsible for

[1] Jacobi, a German who spent almost the whole of his scientific life in his adopted country, Russia, was one of the few men who understood the possibilities of electricity from the start. He invented and actually sailed an electric motorboat on the Neva in 1838. But the atmosphere of autocracy and subordination to foreign capital and science prevented his ideas being realized in his time (see p. 145).

the first installation of electrically-fired sea mines to protect the harbour of Flensburg against the Danish navy. Batteries proved unreliable and too heavy for explosive work and Siemens devised in 1860 a simple armature wound on an H-shaped core, which made the first practical high-voltage *magneto*, useful for blasting and even more for the internal combustion engine's sparking plug of thirty years later.

Wilde in 1867 had the idea of using the current of one Siemens magneto to magnetize the soft iron field magnets of another machine, thus enormously increasing its action and creating the first *dynamo*. He did not, however, develop it further, though he made several machines for electroplating, as he lacked the great resources which the Siemens firm had already built up by its telegraph and cable developments. Siemens adopted the idea of regenerative working[1] and began to make dynamos for extensive dock, railway, and street lighting. He was, however, by no means alone in the field. The Belgian engineer, Gramme, in the early seventies was the foremost in introducing improved forms of armature and in making practical machines with determinable and economic performances. He also discovered that his dynamos, if fed with current, could act as electric motors. Somewhat later, Brush, in America, developed dynamos giving very steady currents and set up the first economical electric street arc-lighting system. The Brush company was also to provide the plant for most of Edison's later incandescent lighting schemes and ultimately to merge into the great monopoly, the General Electric Company, founded in 1892.

[1] Already a familiar idea in the family, cf. pp. 104 f.

Electric Light and Power

By the beginning of the seventies the electric light was attracting popular attention. Still the only practical device was the arc lamp using a heavy current at low voltage. It was certainly susceptible to improvement—within limits. The cheap and slow-burning 'candle' of Jablochkov eliminated all mechanism and was used to light the Boulevard de l'Opera in Paris in 1876.

Swan and Edison

Yet the arc lamp remained clumsy, noisy and far too brilliant for the main use of lighting in the home. The *sub-division of the electric light* into smaller and handier units became a burning problem for the inventors. It was solved practically by two men, Joseph Swan and Thomas Alva Edison, whose careers offer the most violent contrasts.

Swan was born in 1829 in Sunderland and spent most of his life in the neighbouring Newcastle running, with his brother-in-law Mawson, a chemist's shop where he was to stay for forty years. He was a modest, unassuming man, devoted to science as a spare-time hobby throughout most of his life. Mawson and Swan's was no ordinary chemist's shop. Swan had early been interested in amateur photography; he perfected Fox Talbot's and Poitevin's carbon process of printing and made it a commercial success in 1864, thus going into the photographic supply business. Later he was to improve and manufacture dry plates and to invent and produce bromide paper. As an incidental by-product he invented, but did not exploit, chrome tanning. His work on developing the electric light was a spare-time amusement undertaken desultorily as far back as 1848 but taken up with vigour, though still in his spare time, as a result of the public interest in the seventies.

Edison was born in 1847 at Milan on Lake Erie. Of a restless and enterprising disposition, he ran a newspaper in a train at the age of twelve and became a roving telegraph operator at fifteen. He invented several improvements in telegraphy and among them a stock ticker which was most profitable in the era of frenzied finance of 1869. By 1870 he had a works of his own devoted to the exploitation of inventions—the first industrial research laboratory—employing the then-unheard-of number of a hundred workers. From that shop at Menlo Park were to come quadruplex telegraphy, the gramophone, a major part of the telephone and the mechanism of the cinematograph. From it also was to come Edison's contribution to the electric light.

It can be seen that Edison and Swan were very differently equipped to attack the problem of the electric light; Edison approached it from the electrical side, Swan from the chemical. The form of the solution was known in advance to both men. It was necessary to heat a filament by an electric current in a bulb free of air—before the days of the rare gases that meant *in vacuo*. The problems were essentially technical—how to make a filament of high resistance to suit manageable currents and supply lines; how to evacuate the bulb; and how to make and divide the current.

Problems of development of electric light

The first problem had already received many partial solutions and one almost complete one. In 1872, A. N.

Lodygin in Russia had made lamps with short, straight carbon filaments but he lacked facilities for following up the invention and starting commerical production. The first to reach this stage was Swan who solved the problem of forming a carbon filament in an elegant way. He used first carbonized paper, then paper gelatinized with sulphuric acid, finally in 1883, a thread of dissolved cellulose pressed through a die. This process was also to mark the beginning of the artificial silk industry and was actually taken up for this purpose by one of Swan's collaborators, Topham. Edison also used paper and bamboo fibres, as well as some metals, but he was not technically so successful as is shown by his adopting Swan's methods when their joint company was formed in 1883.

The second problem, the evacuation of the bulb, was made possible by the work of the scientific glass-blower, Sprengel, in 1865. He was concerned with producing those scientific toys, the Geissler discharge tubes, first introduced for the analysis of spectra, which were to be the ancestors of all our X-rays and electronic tubes, of fluorescent lighting, and television. The Sprengel pump provided a far higher vacuum than had been attained previously and made it possible to make and seal a bulb with a reasonable life. The technical problem of the production of vacua had stagnated for about two hundred years. The pumps available in the mid-nineteenth century were hardly better than those of Boyle; they had shared in no way in the immense mechanical improvements of the intervening period. Here was another clear case of the law of supply and demand in the development of science and technology. Until some use could be found for high vacua there

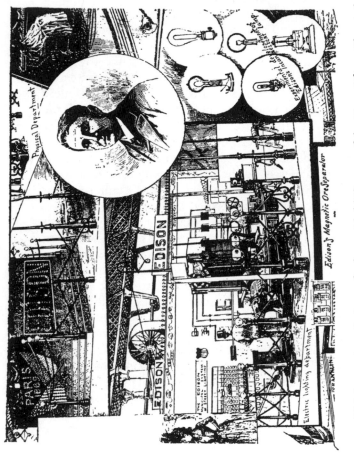

FIG. 5. This is a section of a larger display entitled 'Glimpses of the Great Edison Electrical and Phonographic Exhibit, Paris Centennial Exposition of 1889', which appeared in the *Electrical World* of November 1889. It shows early forms of the incandescent lamp, a power station, and a row of arc lamps.

was no point in developing pumps; without good pumps to produce them, the properties of high vacua could neither be studied nor exploited. The technical difficulties for improving pumps were not great and no new scientific principles were needed. The Töpler and Sprengel pumps were typical laboratory products; cheap, adequate for small-scale work, incredibly tedious to work. Once the large-scale demand for bulbs came in they were rapidly discarded in favour of improved mechanical versions of Boyle's piston pump and later by vapour jet pumps based on the principle of the injector steam pump of a hundred years earlier. These improvements due to technology returned to science in our own time and made possible the high-vacuum technique of electronics and particle accelerators.

The third problem, the power supply, was technically and economically the most important. Here Edison took the lead. His genius at circuit design and his experience at laying out—often illegally—ticker-tape lines, enabled him to imagine the whole complex of mains and distribution lines for district lighting and the creation of central power stations. The first was the Pearl Street station of the Edison Electric Illuminating Company of New York, opened in 1882. Swan had considered electric lighting installation in a more aristocratic country-house way, like the plant he installed in Sir William Armstrong's mansion, driven by a waterfall in the grounds, or in Lord Kelvin's, driven by an oil engine. However, even here he made an important contribution by improving the Planté-Faure accumulators by forming plates with lead oxides in a grid, the method still used seventy years later in every motor-car battery.

Electric power

What Edison had done was to lay the foundation of the age of electric power as a commodity. The provision of electricity for sale involved the use of circuits of a complexity beyond anything telegraphy required; it called for electricity at a steady voltage and it demanded a mass production of measuring instruments to control and meter the current. All this, together with the tasks of installing the service, gave an immense fillip to electrical engineering. Even professors like Elihu Thomson and Edwin J. Houston, who had originally merely tested the performance of electric generators for the Franklin Institute[1] in 1870, soon after set up in manufacturing business as the Thomson-Houston Company.

Now that electricity was worth making centrally for domestic lighting, the current might just as well be used for driving machinery through motors. The large central station serving factories with power as well as lighting was a natural consequence but it took a long time to switch from belt to individual motor drive. At the same time, during the eighties and nineties, Werner Siemens and Schuckert, who had worked for a time with Edison, were developing electricity for traction in trams and railways. By the end of the century electric trams, taking the place of horse-drawn trams, had become a commonplace in all cities, great and small, and remained the main popular means of transport until the advent of the motor omnibus.

It was the development of central electricity stations,

[1] The Franklin Institute, founded in 1844, is the oldest institution in the United States devoted to the study and promotion of the mechanical arts and applied sciences.

with their relatively constant demand for power, that gave a new impetus to another form of electromagnetic generator—the alternating-current dynamo. Its development had followed the same steps as that of the direct-current dynamo and in its history the same names of Siemens, Wilde, and Gramme occur. Batteries, to balance out the load, could not be used with alternating current, but it had the ultimately far more important advantage in that it could be converted from low to high voltages and back again by means of the transformer—essentially a glorified form of Faraday's induction ring of 1831. The transformer as a commercial proposition only came fifty years later, largely through the enterprise of the French engineers, Gaulard and Gibbs. The advantages of high-tension transmission, first for short and then for long distances, were first realized in America by the Westinghouse Company. Soon the superiority of the alternating system was accepted almost universally, despite a stern battle by the advocates of direct current, now led by Edison in the role of a conservative.

Once electric distribution on a large scale was proved to be feasible and immensely profitable, there came a new demand for large efficient power sources, leading to the use of the water turbine first devised by Fourneyron in 1836 and developed for high speeds by Pelton in 1884, and to the development of the impulse steam turbine by de Laval in 1882 and the reaction turbine by Parsons in 1884. The wheel of power production had indeed turned a full circle, back through the reciprocating engines of the eighteenth century to the original aeolopile turbine of Hero of Alexandria.

Electric Light and Power

Once Edison's first station had shown that it could pay, electric power production had a fantastically rapid growth and extension compared to the stumbling progress of the first fifty years from 1831 to 1881. In the twenty years that followed all the essential features of the production and distribution of electricity had been evolved and were in use. From then on the use of electric power has grown at a rate unequalled even in the first Industrial Revolution. This contrast between early slow and late rapid development is a general phenomenon in the application of science to industry, but it has hardly ever been shown so clearly as in the case of electricity. Both in quantity and quality the effort put into the development of electric power production between 1831 and 1881 was small compared either with what went into the development of the telegraph up to 1860, or to electrical engineering between 1881 and 1900. No really great scientific names are associated with it and the total number of effective man-years of work is unlikely to be as much as five thousand.

The logic of development of electrical engineering

It may be claimed that technical difficulties imposed this slow rate, that it was necessary to wait for the Sprengel pump or regenerative winding. That pump, however, was a poor device soon superseded, as we have seen, when a real, technical need arose. Regenerative winding was only a matter of substituting an electromagnet for Faraday's permanent magnet. However simple that substitution was, it may be urged that it is necessary to think of it first and that the production of radically new ideas is the function of genius which cannot be hurried. But the people who did think of it were not geniuses—none could compare with Faraday—and others no more gifted could have hit upon these ideas earlier if the field had attracted enough workers. Here there was not, as there has been in other regions of advance, a technical barrier, such for instance as that imposed by the absence of a light prime mover on aviation. The barrier, or rather the absence of stimulus to advance, was economic. Electricity developed quickly when it paid and not a moment before. Capitalist enterprise had neither the conception to foresee more than a year or so ahead, nor could the money have been found to back such foresight for several years without return. Electricity had to grow on its pickings as it clambered from step to step. What was it that made electricity suddenly profitable in 1881? It was essentially the removal of the obstruction of the limited market already referred to on p. 120. Once a sure demand for current could be established at a profitable price, the generators could be built and improved, more current at a lower price could become available, and a real electrical boom could get under way. It was the incandescent lamp that brought electricity into the vast market of the shop and the private house. That is why the relatively trivial and semi-technical improvements that led to a serviceable electric filament bulb were so vitally important. Its advent marked an absolutely critical step not only in electric technique but in power generation and application. Its use made possible the growth of a new industry, scientific and monopolistic throughout from the start, that took its place with the steel industry and the chemical industry as the main supports of twentieth-century capitalism.

Electric Light and Power

Scientific consequences of the electric bulb

The effects of the electric lamp on science were at least as important as those on industry. Vacuum technique and the discharge tube had existed before the rise of the lamp industry, but they were laboratory crafts requiring much patience, glass-blowing and a lavish use of sealing wax. Indeed it was still with this technique that the great works of the J. J. Thomson and Rutherford period were carried out. The enormous subsequent advances of physics that are associated with the development of electronics and with such gigantic machines as the cyclotron, and leading up to the liberation of atomic energy, would have been impossible without a parallel development of the electrical industry and, in particular, of methods of high-speed evacuation first used in the electrical bulb industry.

At one remove the effect was to be even greater. Edison noticed, in 1883, in the course of a research to determine why his lamps blackened after use, that the hot filaments could retain positive but not negative charge—in modern language, that they gave off electrons. Sir Ambrose Fleming, who worked for a while in the London Edison firm, used this to construct the two-electrode valve in 1904—the first successful device depending on the use of electrons. De Forest, in 1907, added another electrode—the grid—and almost accidentally devised the first electronic amplifying device. Both types of valve found immediate use in the new wireless telegraphy and so, being needed in enormous numbers, could be manufactured cheaply in lamp factories. Coming back to science they provided, as amplifiers and elements in electronic circuits, an indefinitely flexible device without which not only modern physics and chemistry but also biology would be crippled. The electronic industry of the twentieth century with its radar and television is the nursery of modern physics, providing the money and the requirements for professional training that keep physical laboratories full. It owes almost everything to the incentive that led to the development of the first practical electric lamp.

Chapter VI

CONCLUSIONS

THESE samples of nineteenth-century developments, thus summarily presented, should in themselves serve to answer some of the questions about the interaction of science and technique and of their dependence on economic and social factors that were raised at the beginning of this essay. At the same time it must be admitted that in the very richness and variety of the events they record they may convey an impression of confusion. The confusion is real enough, it was even somewhat deliberately provoked at the time as the expression of a natural and unrestrained interplay of initiatives in a period of free enterprise and competition. The nineteenth century was fully aware of itself as an era of Progress. It was then almost impious to examine how that progress came about save as a removal of the restrictions of an earlier and inferior aristocratic domination. But we, to whom all this is past history, must needs strive, if we are to understand and control the developments of our own time, to make some sense out of the maze of events that have led to the world as we know it. The very difficulty of doing so may conceal a clue.

Conclusions

Character of scientific and technical advance in the nineteenth century

What is most apparent in each one of these histories are the contradictions—the easy and rapid realization in practice of some ideas at some times, the many false starts and halting progress of others. Intense and optimistic effort contrasts with unenterprising timidity. In this there is nothing that should surprise us. The developments of science and technique must represent on an exaggerated scale the booms and slumps of contemporary economic life, equally blind in detail, equally inevitable as a conjunction of unstably balanced economic forces. The oscillations in science and techniques must needs be even wilder than those of current production or even of capital investment because the vigour of science depends not so much on the level of techniques as on its rate of change. Not only that, but science is internally a very explosive enterprise. Scientific progress tends to come in bursts because it depends so much on the interaction of many minds. In any period when there are a substantial number of workers in a field and there is an atmosphere of optimism, ideas and application come thick and fast—a kind of chain reaction is set on foot. Contrariwise, if effort in any field falls below a certain level, if workers are isolated or overspecialized, pedantry takes the place of enterprise and the process of discovery comes to a full stop.

To consider only the instability of science would, however, be to give a one-sided picture. Science had in itself, in its laws, in its accumulation of registered facts, an element of inner stability that prevented external economic or social factors from distorting the direction of its progress, although it might speed up or slow down its rate. Another element of stability comes from the fact that, already in the nineteenth and even more in the twentieth century, progress, in spite of its ups and downs, has been so rapid that single individuals may live and work through several of its oscillations and thus preserve a continuity of purpose and even of plan. Faraday and Pasteur were both men who had a clear vision of what they intended to do and for the most part achieved it unless they hit on something better on the way.

In discussing the progress of science, both its instability and its stability have to be taken into account, as well as the external and internal factors that serve to determine its course and speed. I will attempt in the first place to indicate the factors in the nineteenth century that operated to drive science and technology on to achieve their triumphs in the understanding and control of nature. I will then turn to those that held up this progress and prevented the scientists and inventors from attaining more than a small fraction of what was within their reach. Thirdly, it will be necessary to ask why the great triumphs of science and technology in the nineteenth century brought relatively little benefit to those who helped them forward and why the world at the end of the period was so evidently in greater peril and anxiety than at the beginning.

Positive factors. The driving force of profit

The principal driving force behind the uneven but repeatedly accelerated progress in science and technology in the nineteenth century was the enormously increased market for industrial products as their costs fell and the profits on them increased. This in turn was a consequence of the quantitative spread of the effects of the qualitative change brought about by the Industrial Revolution in the eighteenth century. The more kinds of things that were made by the new methods, the more openings there were for inventions and, at one remove, for the science on which they were based. This in turn led to a demand for more science teaching, which provided the necessary sustenance for the growth of academic science.

In other less material ways the intellectual atmosphere of the nineteenth century favoured the adventure of science. It was the ideology of the rising class of manufacturers, though they were apt to drop the ideal of scientific progress when they had made their pile and moved into high society. It was a climate of mind predominantly liberal, progressive and anti-clerical. Outside Britain and France, where it had been more or less assimilated by the ruling classes, it also had a revolutionary aspect, as in Russia, or a nationalist one, as in Poland and Italy.

The spread of science and technology

The nineteenth century witnessed the beginning of a significant spread and shift in the centre of gravity of science. At its outset it was limited to the area of cultivated society of the eighteenth century. This centred on France and Britain, which set the tone, but included small and select groups of liberals in the countries touched by the Enlightenment—the Low Countries, the German courts, Switzerland and Italy. Subsidiary centres were established in Scandinavia, in Russia, and on the eastern seaboard of the United States.

In the first decades of the nineteenth century, supremacy in science seemed to have passed definitely to France. The great new institutions like the École Polytechnique and the École Normale, which had been born in the Revolution, had served to bring science for the first time into the official state framework. They survived the patronage of Napoleon and even for a while the suspicion and neglect of the restoration.

Science in Britain. Stagnation and recovery

During the same period science in Britain had been partially under a cloud on account of its suspected alliance with atheism and sedition. The great outburst of optimistic utilitarian science of Franklin and Priestley was checked, but only for a while, and with the exception of the Royal Institution it never received official favour. Nevertheless the Industrial Revolution went on its triumphal way as a primarily British movement and the demands of the new industries on science ensured its revival.[1] Once again in the twenties and thirties a new wave of scientific advance began, and one with a definitely radical flavour. The new sponsors of science, including such dissimilar characters as Bentham and Brougham, Birk-

[1] The revival was a very uneven one. The physical sciences came first with physics, particularly electricity, in the van and chemistry at the tail till the middle of the century. These were the sciences that naturally raised least apprehension on religious and political grounds. Biology and particularly geology had a much harder fight for it. See pp. 163 f.

Conclusions

beck and Babbage, despaired of the reform of old institutions and created new ones—the London Mechanics'
Institution in 1823 (later Birkbeck College) and London
University College (the godless college in Gower Street)
in 1827. Their greatest achievement, however, was the
foundation of the British Association.

The British Association for the Advancement of Science

Here the leading spirit was Charles Babbage, who had
become, with the publication of *The Decline of Science*[1] in
1830, the arch-critic of the Royal Society, largely for its
failure to live up to its seventeenth-century aim 'To
improve the knowledge of naturall things, and all useful
Arts, Manufactures and Mechanick practises, Engynes
and Inventions by Experiments'.[2] Babbage drew his
inspiration from the Deutscher Naturforscher Versammlung, founded in 1822 by that unfortunate but courageous
pioneer, Lorenz Oken. The British Association for the
Advancement of Science met first at York in 1831,[3] the

[1] *Reflections on the Decline of Science and on Some of its Causes*, London,
1830.
[2] C. R. Weld, *A History of the Royal Society*, London, 1848, Vol. I,
p. 146.
[3] A most optimistic tone characterized the opening address by the
Rev. William Harcourt, its first general secretary, though it carried
undertones of protest against retarding social forces outside and inside
the body of science itself:
'I do not rest my opinion, gentlemen, of this want upon any com-
plaint of the decline of science in England. It would be a strange
anomaly if the science of the nation were declining, whilst the general
intelligence and prosperity increase. There is good reason, indeed,
to regret that it does not make more rapid progress in so favourable

year before the Reform Bill. Its deliberate aim was 'To give a stronger impulse and more systematic direction to scientific inquiry, to obtain a greater degree of national

a soil, and that its cultivation is not proportionate to the advantages which this country affords, and the immunity from vulgar cares which a mature state of social refinement implies. But, in no other than this relative sense, can I admit science to have declined in England . . .' (p. 17).

'I am not aware, gentlemen, that in executing such a plan we should intrude upon the province of any other institution. There is no society at present existing among us which undertakes to lend any guidance to the individual efforts of its members, and there is none, perhaps, which can undertake it. Consider the differences, gentlemen, between the limited circle of any of our scientific councils, or even the annual meetings of our Societies, and a meeting at which all the sciences of these kingdoms should be convened, which should be attended, as this first meeting you see already promises, by deputations from every other Society and in which foreign talent and character should be tempted to mingle with our own. With what a momentum would such an Association urge on its purpose! What activity would it be capable of exciting! How powerfully would it attract and stimulate those minds which either thirst for reputation or rejoice in the light and sunshine of truth!

'The Royal Society still embodies in its list every name which stands high in British science; it still communicates to the world the most important of our discoveries, it still crowns with the most coveted honours the ambition of successful talent, and when the public service requires the aid of philosophy, it still renders to the nation the ablest assistance and the soundest counsel. Nevertheless, it must be admitted, gentlemen, that the Royal Society no longer performs the part of promoting natural knowledge by any such exertions as those which we now propose to revive. As a body, it scarcely labours itself, and does not attempt to guide the labours of others . . .' (pp. 18–19).

O. J. R. Howarth, *The British Association:
A Retrospect*, London, 1931.

attention to the objects of science, and a removal of those disadvantages which impede its progress, and to promote the intercourse of the cultivators of science with one another, and with foreign philosophers'.[1]

The Association rapidly became and remained for the rest of the century the spearhead of scientific advance in Britain. Its annual meeting became the favourite battle-ground of ideas, rising to their peak in the Thomas Henry Huxley—Bishop Wilberforce clash on evolution at the Oxford meeting of 1859. It took the place, through its committees, of extremely *laisser-faire* governments, in such matters of national concern as the Magnetic Observatory at Kew, the development of meteorological instruments, the calculation of mathematical tables, the establishment of electrical standards, and it even helped to finance research and scientific expeditions in these and other fields.[2]

[1] *Ibid.*, pp. 16–17.

[2] The details of these grants furnish invaluable material on the quantitative effort on science in nineteenth-century Britain, at that time the most developed and wealthy country in the world. They amounted in all, over the whole first century of its existence, to about £92,000, less than £1,000 a year on the average, though the actual rate was at its greatest in the middle of the period. Of this sum 47 per cent went to the physical sciences, much of it to magnetism and meteorology; 35 per cent to the biological sciences, including geology; and 18 per cent to the social sciences, largely to finance anthropological and archaeological expeditions (for details, *ibid.*, pp. 266–92). It is clear that such sums, even allowing for the small scale of the effort and the greater value of money, could not have *paid* for the researches in question. They served rather to cover some of the out-of-pocket expenses of investigators of private means and, by giving them a seal of respectability, to encourage donations from the wealthy.

Conclusions

The phenomenal wealth of the British middle classes throughout the nineteenth century led to the development of a highly individualistic science organized seriously, but somewhat chaotically, by men of science themselves, including an unprecedentedly large number of wealthy amateurs.[1] This was at first a strength, but later a weakness of British science. The concentration of power in industry and government at the end of the century left little place for the independence and enterprise that had built it up.

A corresponding development in mid-nineteenth-century France could not take place for lack of industrial backing. Individual achievement was on as high a level as ever. The story of Pasteur brings out both the opportunities for science in France and the stinginess and neglect to which it was subjected.

The German hegemony in science

By the last quarter of the century leadership in science had passed, at least quantitatively, to Germany. There up to 1848 science, despite highly circumscribed official patronage, had formed part of the liberal movement of the Enlightenment. The somewhat mystical and absurd prophet of *Naturphilosophie*, Oken, a friend of Goethe, had, as we have seen, founded the Deutscher Naturforscher Versammlung, largely to emphasize the liberal aims of science. In 1819 he resigned his chair rather than censor

[1] Among those who contributed notably to science and who were able to equip important laboratories from their own resources were Carlisle, Joule, Wollaston, Young, Sturgeon, Rayleigh, Gassiot, Grove, Sorby, Nasmyth.

his scientific political magazine, *Isis*. In Germany the forward movement of science owed much to the growth of a new spirit in the universities. While in France and England, though not in Scotland, the older universities had carried out a successful rearguard action against the advance of natural science till well into the twentieth century, those of Germany had owed their rise from a low point of eighteenth-century torpor first to the Enlightenment and then to nationalist feeling. Paradoxically, it was the new university of Göttingen, founded by the half-English king George II in 1736, that led the way. The Elector could do what he liked in his own dominions but could not touch the obscurantist prerogatives of Oxford and Cambridge.[1] Numerous and well-supported, the German universities provided for the best part of a century by far the largest opportunities for education and research in the world. And it was not for Germans alone. Americans, Englishmen, and even an occasional Frenchman, studied science in German universities, and it was there too that the now universal higher degree of Doctor of Philosophy first became popular. The liberal tendency of science in Germany did not survive the impact of the Industrial Revolution.

[1] On the death of William IV the Kingdom of Hanover passed under Salic law to Victoria's uncle, the Duke of Cumberland, who proceeded at once to abrogate the constitution and with it the privileges of the University. This was so resented that it led to the resignation of the most distinguished professors, the famous Göttingen Sieben.

Though Göttingen later recovered much of its academic liberty it was in the latter part of the nineteenth century somewhat overshadowed by Berlin.

Conclusions

It was after 1848, with the compact between the bourgeoisie and the princes, that the scientists became pillars of state and scientific education was favoured and magnificent laboratories began to be built. It was from Germany that the next phase of science was to come, with its close links with the new monopoly industries, particularly the chemical industry, and with the state, particularly in its military aspect. The history of the Siemens family brings this out most clearly.

The predominance of science was secured over a still wider field by the rapid expansion of standard scientific publications. German *Zeitschriften* and *Handbücher* covered every aspect of science and made German an almost obligatory scientific language. In consequence German science, and often German professors as well, virtually colonized the scientific world outside the ancient centres of Britain, France and Italy.

The smaller European countries, with their old and high tradition of scientific work, tended to fall largely in the German orbit, though France had an influence in Belgium and Switzerland, and Britain in Scandinavia. In these countries there was no place for an autonomous scientific tradition but their contribution, as witness the names of Berzelius, Van't Hoff, Naegli, Solvay and Nobel, was out of all proportion to their size. Italy also, throughout most of the century, was too occupied with the effort of national liberation to contribute her share, though Avogadro and his brilliant expositor Cannizzaro took the decisive step in establishing the reality of the theory of atoms by showing how they could be counted.

Conclusions

Science in Russia

The development of science in Russia was as characteristic of its native genius as of its political and economic backwardness. At the beginning of the century, science was well-established there in the academies and the universities, due largely to the initiative and enthusiasm of Lomonosov who ranks with Leibniz in Germany and Franklin in America among the great founder figures of national culture. All through the century came a succession of brilliant individual scientists, Jacobi, Lobachevski, Lenz, Jablochkov, Mendeleev, Butlerov, Metchnikov, all of whom made important contributions to world science, together with many others whose names are hardly known outside Russia. The advance of science inside the country was, however, doubly hampered by the autocratic government and the essentially feudal social system. Tradition combined with fear of the revolutionary implications of science led the government to favour foreign rather than native science and to fill the academy largely with German scientists. As industry was also largely in the hands of foreign concessionaires the numerous inventions of Russian scientists were not taken up, so that the potential contribution of Russia to technological advance was not realized during the nineteenth century. Only towards its end, the growth of a native capitalism provided conditions for a science free from foreign domination, but this was only to bear fruit during the next century.

Conclusions

The United States. Inventor's paradise

In the United States the conditions were very different. The original scientific impetus of Franklin and Jefferson had largely disappeared by the nineteenth century. The extremely *laisser-faire* atmosphere of Jacksonian democracy did not favour federal or state aid to science, and the colleges and universities were on the whole conservative institutions which looked to Europe for inspiration. The United States did produce some of the most eminent men in the physical sciences in the nineteenth century, such as Henry and Gibbs, but they had little impact in their own country. The reason was that, in a continent being rapidly opened up with a growing and shifting population, there was no place for the European type of intellectual, nursed in a highly traditional and stratified society.

It was quite otherwise in the technical field. There the conditions were peculiarly favourable for ingenuity. Great resources, shortage of labour, long distances, all put premiums on machinery and the greatest degree of automatic working. The inventor, who needed neither schooling nor capital, had an open field. Agricultural machines, sewing- and bootmaking machines, typewriters, revolvers are all labour-saving inventions inspired and fostered by American conditions. Even more significant for the future were the laying of the foundations of the fabrication of machinery from interchangeable parts, first developed for small arms by Eli Whitney, and that of the assembly line, which first appeared in the slaughterhouses of Cincinnati. The coming together of these methods was to engender the mass production of the twentieth century.

Conclusions

There was nothing like the same need for capital-saving inventions. Here the major advances, in steel and chemicals, were made in Europe. American economy could afford the waste of materials and indeed encouraged it.

Even at the end of the century, as the story of Edison shows, there was still far less contact between the professional scientist and the inventor than in Europe.[1] Inventors found plenty of scope in the application of a few simple and old scientific ideas, while scientists remained almost unaffected by the wealth of constructive devices that were coming into use all round them.

Retarding factors. The failures of science

This general survey of the atmosphere in which science developed throughout the nineteenth century shows in general how favourable it was for its advance and, though to a lesser degree, to the speed and fruitfulness of its applications. The other side of the picture still remains to be explained. How was it that the advance of science in one field after another was so halting and uneven? The march of history, in science as in other fields of human enterprise, needs in the first place to be correctly traced. The achievements of science need to be praised as examples to future generations, but so also do its failings. To detect the latter, however, a far more critical approach is needed than to expound the former. The achievements of science are intrinsically admirable. Any trained scientist can appreciate, in reading the works of his predecessors, the elegance of the methods they used and their success in

[1] There were, however, notable exceptions such as those mentioned on p. 128.

overcoming ancient obstacles. The achievement may yet lack a full measure of appreciation, since, with our present knowledge, we may have forgotten or underestimated these obstacles, but the need of praise remains.

The failures of science, however, go most often unrecognized. Usually they cannot be attributed to any particular scientist but apply to a school or a generation of science. They are, however, for that very reason more rewarding to search out than the successes. Successes are noted and copied, failures go unrecorded and their causes still remain with us and hold up the science of the present and the future. To detect them more than a study of separate branches of science or of technology is necessary. The failures need not be inside the discipline or industry itself, they are as likely to lie in the absence of contact with ideas and practices from outside. In science they may come from lack of contact with industry and vice versa. The examples I have given in the more detailed part of this essay have shown some of the main types of failure both in the advance of science itself and in its applications.

I have shown there something of the relations existing between the development of fundamental ideas, such as the conservation of energy, or of whole fields of science, such as electromagnetism, and the development of the technical devices such as the steam engine, the telegraph, and the electric light. These were so close and intricate that it is difficult to disentangle the support and inspiration that the technical gave to the scientific side from the value of the precise knowledge and new phenomena that the scientific gave to the technical. However we may resolve this riddle, it cannot be denied by anyone acquainted with the facts that the rate of progress of either branch depended on the other so much so that either stagnation in one or the other or imperfect contact between them held up the advance of both.

Physical science all through the nineteenth century was dependent, entirely for support and largely for inspiration, on industry—directly, by the financing of scientific research and education by industrialists; indirectly, by the building up of independent fortunes which, though in very few cases, could be used to support amateur researches. To a much lesser extent was industry dependent on science. The results of the science of the past could be embodied in such essential devices as the steam engine, but once in practical hands the need for current science for immediate further development was small. It grew, as we have seen, throughout the century, and by its end the new, almost purely scientific industries of electricity and chemistry foreshadowed an age when not only the progress but the very day-to-day existence of industry would depend on science.

The limited impetus of industrial advance

The relative independence of industry at the beginning of the century goes some way to explain the little help it afforded to science, and the slowness with which it assimilated scientific ideas. Where discoveries opened new fields, as in the cases of electromagnetism and organic chemistry, there were long delays in following them up and this only occurred when there was an overwhelming case for profitable exploitation. We have shown by example some of the basic reasons for this delay. They are in keeping with the general character of a rapidly growing competitive capitalism. This growth, with its ever-increasing markets opened up by rail and steamship, put a premium on the multiplication of existing production rather than on technical advance. The key industry—textiles—showed no fundamental improvement throughout the whole century.

Only where the very quantitative increase in the production of goods led to new demands, such as those for rapid transport and communication, was there a demand for radically new solutions. Even in these cases, however, the opportunities for introducing technical changes were limited. In boom times markets were assured without any need for innovation; in slumps new investment was unthinkable. Only in periods of recovery was it worth while thinking of them. The capitalists were willing to accept an obvious innovation of proved workability, when it was offered from outside, but rarely to put up money of their own for its invention. This is well brought out in the case of the development of steel, where not one of the major improvements came from inside the industry itself.

Breaches between science and industry

Nor were the scientists, in general, eager to intervene in industry. Throughout the century there was a growing separation of the scientists from the manufacturers. The intimate, personal and family connections that had existed between science and industry in the later eighteenth century gradually diminished as the new century wore on and found their echo only in the annual beanfeasts of the

British Association. This separation was quite as much due to the success of science as to that of industry. As the century progressed, science began to play a larger and larger part in the universities and government teaching establishments, first in France, then in Germany and Britain, lastly in Russia and the United States, to name only major countries. The talented amateurs, who in the earlier part of the century practically monopolized science except for a tiny élite of academicians or, in England, of beneficed clergy and college fellows, gave way to the professors and by the end of the century could no longer compete with them in scientific discovery.

It was otherwise, as we have seen, in industrial advance. Even at the end of the century the major innovations were still coming from a race of inventors without a university background, who had learned what little science they needed from books and from their experience in workshops and laboratories equipped with their own hands. The great success of inventors like Edison was, however, to presage a new phase, what might be called the industrialization of invention, with the setting-up of large research laboratories. The industrial laboratory and the government research laboratory that came with it brought science back into industry in a new way. The consulting scientist and the scientific entrepreneur were gradually replaced by the whole-time salaried scientist and the new profession of scientific research worker was created.

The training of scientists and technologists

These changes, though they were brought to completion only in the twentieth century, were visible as tendencies towards the greater organization of and a recognized status for science and technology all through the nineteenth. They brought with them advantages and disadvantages. At the beginning of the century, the absence of any provision for training and finance was compensated by the relatively simple nature of applicable science and the opportunities for the amateur and the inventor to be accepted into the upper ranks of science and industry. Before the century was over the facilities for learning science in industrialized countries, particularly the newly industrialized Germany, were becoming comparable to those of the older learned professions and scientists were winning a recognized place in the direction of industry. For these very reasons the entry to science and technology tended to be more and more limited to the intelligentsia of the minor bourgeoisie and it became almost impossible for the lone inventor to succeed unless he came to terms, usually rather unfavourable terms, with the big firms. There were to be no more Gilchrist Thomases.

The divorce of science from industry in the nineteenth century was, however, as I have shown in the more detailed studies, a partial and relative one. In certain fields, notably chemistry, it hardly occurred and over the whole of science one can only say that the relation was less close than it had been in the eighteenth or was to be in the twentieth century. Nevertheless, it is possible to see in case after case how it served to retard the progress of both industry and science. This divorce was not an isolated phenomenon, it was associated with other retarding factors inside the structure of science itself; which in turn were deeply affected by economic and other social forces.

Conclusions

The nineteenth century, as was pointed out at the beginning of this essay, was predominantly the century of transition for science. It was to change during those hundred years from an elegant ornament of society, practised by virtuosi, to an essential factor in the everyday production of goods and services. Science had in that time to fight its way into education and the professions. The process inevitably diverted much of its energy and the compromises it was forced to make necessarily weakened its capacities for advance.

The battle for scientific education. Revolution, reaction and reform

The battle for scientific education and for the place of science in general education raged all through the century. Educational institutions are normally conservative and to find place for science in a curriculum already crammed with the relics of renaissance humanism would have been difficult enough. But in the beginning of the nineteenth century another factor powerfully reinforced the resistance. Before 1788 science had been all the rage in the circles of the Enlightenment. Even if Oxford and Cambridge had been left 'plunged in their dogmatic slumbers', the dissenting academies in England and the rejuvenated universities of Scotland had provided an education which balanced science and liberal sentiments. Similar movements took place in the enlightened despotisms from Spain to Russia. The French Revolution changed all that. It now appeared that science, glorified by the Encyclopaedists and in which Voltaire himself had dabbled, not to mention the terrible Dr. Priestley, was a dangerous revolutionary doctrine, subversive to church and king. Its spread, especially to the middle or even worse to the lower classes, needed to be sternly resisted. A similar reaction took place in France after the fall of Napoleon and throughout Europe under the Holy Alliance. Even in America, Franklin was forgotten and Priestley died in unhappy exile.

The reaction was not maintained—science came back with the reviving radicalism of the thirties, but the clergy and their landed and brewing supporters were well entrenched in the teaching profession and the battle went on for another sixty years at least, as witness the great evolution controversy. By the time it was over, both protagonists emerged very changed from their primitive state. Science, in particular, was very considerably denatured, with its conclusions watered down by agnosticism and its field of operation rigidly circumscribed to leave full play for a now spiritualized dogma, with less insistence on the literal interpretation of the first chapter of Genesis. Here, however, we are less concerned with this aspect than with the practical result that the facilities for learning any science at all remained very restricted throughout most of the century. What was taught in the older universities before their reform in 1877 was extremely formal and a young man who was interested in science had to pick it up as best he could. The new London colleges[1] and the older Scottish ones gave more place to science but it was still largely theoretical. No practical

[1] For a most illuminating account of their history see *University Reform in London*, by L. T. Humberstone, as also his article on education in *London and the Advancement of Science*, 1931.

science teaching was given in Britain, even in Cambridge, before 1845, and then it grew very slowly.

The College of Chemistry—the germ of the later Royal College of Science and of Imperial College—founded in 1845 at the instance of the Prince Consort was an isolated exception. It was found necessary to import a director from Germany, A. W. von Hoffman, one of Liebig's bright young men, and the brilliance of his pupils such as Crookes and Perkin testified to the need for such an institute. Perkin discovered the first aniline dye and laid the foundation of an industry that was soon to be lost to Germany.

Limited as were the opportunities for scientific training, those for research were even more so. France owed her scientific pre-eminence in the early nineteenth century to the great teaching centres of the Polytechnique and the École Normale, but even there practical teaching was limited to specially favoured assistants of the professor. In Britain, research facilities hardly existed at all. The historian of science is inevitably struck by the unbroken sequence of important discoveries that, from 1797 to 1850 and beyond, came from the Royal Institution, but this was in fact thanks to the foresight of Count Rumford who established what was practically the only research laboratory in Britain which could accordingly call on the very best talent available. Outside it there was little hope of practical assistance except in the cases of wealthy men like young Joule who could build themselves laboratories. Professors had to manage as best they could in cellars.[1] The era of big research laboratories only began in

[1] "In my time," said the late Professor Ayrton, who was a student

the sixties, inspired largely by Liebig, and spread from Germany very slowly all over the world.[1]

Nineteenth-century expenditure on research

It would require the most painstaking research to determine the actual amounts spent on research all through the nineteenth century. The inquiry would be well worth making for it would reveal how far the factor of finance acted in stimulating or retarding the advance of different branches of science at different times. Figures readily

in the 'sixties, "Thomson's laboratory consisted of one room and the adjoining coal-cellar, the latter being the birth-place of the siphon recorder. . . . There was no special apparatus for students' use in the laboratory, no contrivances such as would to-day be found in any polytechnic, no laboratory course, no special hours for students to attend, no assistants to advise or explain, no marks given for laboratory work, no workshop, and even no fee to be paid. But the six or eight students who worked in that laboratory felt that the *entrée* was a great privilege. . . . Thomson's students experimented in his one room and the adjoining coal-cellar, in spite of the atmosphere of coal dust, which settled on everything, produced by a boy coming periodically to shovel up coal for the fires. If for some test a student wanted a resistance coil, or a Wheatstone's bridge, he had to find some wire, wind the coil, and adjust it for himself. It is difficult to make the electrical student of to-day realize what were the difficulties, but what were also the splendid compensating advantages of the electrical students under Thomson in the 'sixties. . . . But oh! the delight of those days! Would we have exchanged them, had the choice been given us, for days passed in the most perfectly designed laboratory of the twentieth century without him? No! for the inspiration of our lives would have been wanting."'

S. P. Thompson: *The Life of Lord Kelvin*,
Vol. I, 1910, p. 297, footnote 1.

[1] See Pasteur's article quoted on pp. 86 f.

156

available show that public support for research in Britain, though on a rising scale, was very modest. The figures quoted already for the British Association bring this out. Of the total of £92,000, £62,000 was expended by 1900. If we add to these those expended between 1800 and 1900 by the Royal Society, £140,000, and £660,000 by the government, mostly for specific projects of naval or military interest, we arrive at a figure of £862,000. This must include nearly all that was spent, outside private laboratories, for the scientific societies lacked funds and the universities contributed very little. Even allowing for the change in the value of money this represents only one twenty-fifth of what is being spent annually on civilian scientific research in Britain today. Corresponding figures from other countries would almost certainly be smaller except possibly for Germany in the latter part of the century. With science teaching on such a small scale and with the even more restricted opportunities for research, the achievements of the nineteenth century in science stand out in their full measure, while the delays and hesitations in arriving at them become more understandable.

The great contrast between the character of scientific and technical advance throughout most of the nineteenth century, which has already been discussed, explains why there was relatively little technical education of a formal kind, its part being taken by apprenticeship, and virtually no scope for industrial research except at the very end of the century. In the eighteenth and early nineteenth century much engineering research had been carried out in Britain by engineers such as Smeaton, Watt, and Stephenson and ironmasters such as Wilkinson and Nasmyth, who paid for it out of their own pockets. In France the state, particularly after the revolution, had also subsidized such research.

Research or sound practice. The professional engineers

As the century went on the expense of engineering research grew. But the engineer was now no longer an entrepreneur operating for his own profits while the owners or the joint stock companies, who replaced them, were not so often technical men and looked on research as pure waste of money. Relative to the needs and knowledge available there can be no doubt that the quality of industrial research declined and did not recover till the twentieth century. The exceptional cases, such as those of the Siemens brothers and of Edison, already discussed, serve to prove the rule. In Britain, Bessemer, for all his ingenuity and enterprise, was only an amateur scientist and as a practical inventor he was not of the calibre of the great engineers of the early century. With the growth of machine industry there came a re-establishment of technical traditions and of a great body of men, regarding themselves as part of a new profession of engineers and no longer as applied scientists. They absorbed as much science as was needed for the use of the drawing-board, the slide rule, and the gauge; they used formulae from standard books. But like the doctors before them their real reliance was on practical experience. What worked once would work again; true it might be improved, but it was unsound and hazardous to venture in the search of radically new principles or to attempt to apply them. Even if they worked how could one tell that they would pay? The fate of great imaginative engineers like Brunel, showed it was safer to stick to established ways. The engineer was also inevitably involved in questions of production and, though rarely as a principal, in those of investment and finance. It was more and more difficult to maintain the disinterested philanthropy of a man of science. Men of conscience, like Faraday, saw early the danger this presented to the pursuit of knowledge. Science and engineering as they grew tended to drift apart.

The gradually increasing incoherence of the scientific and technical activities in the nineteenth century, itself brought about by an enormously increased scale of operation, was, as we have seen, only partly relieved by improvements in education and training. The retarding action due to such failure of communication was essentially of a negative kind. It acted by diminishing the potential chances and conjunctions needed for the appearance of new ideas and by lessening the opportunities for their successful use in practice. Some examples of these have already been given, for instance, in the slowness of the recognition of the principle of the conservation of energy through the lack of contact between industry and science; or of the fifty years' gap between Faraday's electromagnetic discoveries and their full utilization through the lack of contact between science and industry.

These factors, together with the basic lack of funds, account for only a part of the retardation of scientific and technical progress. By the very fact that they operated through a failure or absence it was extremely difficult to recognize them at the time. Besides them there existed, both in science and technology, other powerful retarding factors operating consciously and positively. These were the old beliefs that were used to rule out of consideration, to ban, or at least to discourage, new ideas. No part of science was immune from them, though they were more pernicious in geology and biology than in physics and chemistry.

The retarding effects of tradition in science. The Newtonian stranglehold

In general the effect of tradition in science, as in other fields of endeavour, was conservative. New ideas threaten the prestige, even the competence, of the old who, however keen-minded, have neither the time nor the necessary background to understand them. Until the nineteenth century, except in the limited sphere of mechanics and astronomy where Newton had long reigned supreme, the main objections to new scientific ideas had come from the outside, from the vested interests of the theologians and philosophers. Now, though these still continued the fight, further opposition was generated inside science itself. There had been time to build up a new orthodoxy, indeed several successive ones.

The very comprehensive nature of scientific generalizations made their defenders sensitive to attack at any point. In the physical sciences it was the great tradition of Newton himself that was to be the greatest fetter on advance. Enshrined as it had become in the teaching of Cambridge and in the Council of the Royal Society, it dominated the physical outlook, determining at once which problems were deemed to be important and the proper mathematical way in which they should be studied. Newton's mechanical, mathematical universe was defended most fiercely on unessential points for fear of throwing nature back into chaos. Thomas Young's most persuasively worded advocacy of the wave theory of light led to virulent abuse from which he could not guard himself even by showing that Newton himself might well have accepted it. As has already been shown, the Newtonian predilection for particle dynamics distracted attention from the development of the mass and statistical phenomena of thermodynamics. The same prejudice was responsible for the actual rejection of Joule's epoch-making paper on the mechanical equivalent of heat by the Royal Society.

It was characteristic that the new ideas could not penetrate from the centres of physical science in Cambridge or Paris but had to come from Germany, Scotland, or from Manchester—'where they ate their dinner in the middle of the day'. In the thirties Babbage had campaigned successfully against the 'dottage of Newton' (an allusion to Newton's convenient but restricted symbolism of $\dot{x}$ and $\ddot{x}$) and in favour of the 'd-ism of Leibniz' (the now universal $\dfrac{dx}{dt}$, $\dfrac{d^2x}{dt^2}$) but the battle was only partly won. William Thomson's struggle to emancipate himself from older views showed that even by the mid-century the Cambridge school was firmly wedded to the past. Even Maxwell, who effected the full reconciliation between Newtonianism and field theories, had great difficulty in making himself understood to the physicists of the time. In Germany the comprehensive and deliberately qualitative *Naturphilosophie*, sanctified by Goethe, was as staunchly defended. Yet once it was overthrown in the forties, the reaction against it was so strong that even such a mathematical theory as the conservation of energy could be rejected, as we have seen (p. 60), on the grounds that it opened the way to a mysterious generalized force or energy in the old alchemical tradition.

This resistance to the new in science cannot be explained merely by reference to common human instincts, or to the traditions of science itself. Both the innovations and the conservative tendencies reflected, as battles of ideas and sometimes quite consciously, the protagonists of contemporary political, national, and class struggles. Examples of this have been given already in the case of Pasteur and spontaneous generation, where the success of purely negative experiments could be used to support a transcendental view of special creation.

Anti-atomism

Another traditional bar to the advance of science was the persistent Aristotelian notion of the continuity of substance and with it the resistance to the older atomic theories of Democritus and Lucretius, which were considered intrinsically subversive. It was far safer to accept the cautious view that chemical formulae were a convenient shorthand for explaining the course of reactions, which fitted in with the essentially conformist philosophies of Kant and Mach. Here even Newton's authority found it difficult to prevail. John Dalton, who got his atomism from Newton, is often thought to have established the atomic theory in its modern form at the very beginning of the nineteenth century. Nevertheless, while it was formally accepted and its convenience recognized, its reality was always disputed. It was this flight from the atom that was responsible for Faraday's not pressing the atomic nature of electricity in the thirties,[1] which held back the discovery of the electron by sixty years. It was the same resistance that, as I have shown (pp. 209 f.), held up the logical deduction of the spatial arrangements of atoms in molecules from the time of Pasteur's discovery of asymmetry of 1848, to Van't Hoff's hypothesis of stereochemistry in 1874, and it was not in fact till after the end of the century that the material atom was finally and reluctantly admitted by all the chemists.

Genesis and geology

In the fields of geology and biology an even earlier tradition had to be combated: the literal interpretation of the book of Genesis, abandoned by the Pope himself only in 1951. This was in fact the battle-ground *par excellence* of science and religion. It was that and much more. It was a clear demonstration of how political reaction could arrest and turn back the course of science for decades. Here it is pertinent to remark that the slowing of scientific advance was not so much due to the attacks on scientific conclusions from churchmen and others outside science, but rather from the reluctance of the scientists, in the face of what now seems overwhelming evidence, to think along unorthodox lines. An examination of the history of thought shows that this was even more a political and class-determined reluctance than one due to piety, for the degree of reliance on the scriptures waxed and waned in

[1] This, as Sir Harold Hartley has shown, was very largely due to the influence of Davy.

harmony with the prevalence of reactionary or liberal parties in the state. The liberal movement of the late eighteenth century, if it had continued unchecked, would have been ready to acclaim the long geological periods of Hutton or the evolutionism of Erasmus Darwin or Lamarck. They showed daring speculation, an exaltation of the forces of Nature and a thorough disrespect for tradition and dogma. As such, they were anathema to the generation of anti-Jacobin and Restoration scientists who sought to please their patrons by showing that science could be useful without being subversive. Many nine-teenth-century scientists were acutely conscious of the danger that the prevalence of infidel ideas, even in science, might present to the social order. It was these considera-tions that led Whewell for instance to ban Darwin's *Origin of Species* from the shelves of Trinity College Library.

In the course of time these fears abated and the success of the reform movement gave new supporters to scientists like Lyell, who showed that the understanding of the rocks, so essential to the railway engineer and the miner, implied some relaxation of the literal interpretation of the story of the Flood.[1] This was only a partial success. By the middle of the century, money, manufacture, and reli-gion were well reconciled and took very ill the shock administered, though with the greatest diffidence, by Charles Darwin. The passions of that controversy were felt, and rightly felt, to be justified because it involved the whole of society and established institutions.

[1] See J. W. Judd, *The Coming of Evolution*, 1910, Gillispie, *op. cit.*, and, for a more literary presentation, the novels of T. L. Peacock.

Conclusions

Evolution and progress

In some ways the theory of evolution suited aggressive and competitive capitalism very well; indeed it was to a large extent created in its image. The struggle for existence and the survival of the fittest reflected free competition and personal success. Nevertheless, the general reaction of the middle classes was that it was a dangerous doctrine. The old liberal view was that the establishment of capitalism marked the achievement of the natural state of society. The prejudices of barbarous times had been overthrown and now that iron laws of economics had been understood there was no sense in any further change. Progress had achieved its essential aim, only a few finishing touches remained to be given.[1] The idea of evolution threatened this complacency of capitalist society. If it had taken the place of other forms, another form of society might take its place. Evolution, for all its respectable protagonists, smelt of socialism and was resisted accordingly.

It is difficult to make any adequate estimate of the degree to which anti-scientific tendencies, outside and inside science, hampered its advance in the nineteenth century, but the effect must have been considerable. Certainly, leading ideas, such as the conservation of energy,

[1] Mid-nineteenth-century literature bears witness to the spiritual and aesthetic difficulties this acceptance of capitalism involved. Witness particularly Tennyson in *In Memoriam*, with its horror at the annihilation implied by 'nature red in tooth and claw', and Dickens' *Hard Times* with its grim picture of the Gradgrind school of practical science.

spatial molecular structure, the continuity of geological change and organic evolution, were held up largely due to these factors for periods ranging from twenty to seventy years. Further, a relatively enormous amount of mis-guided effort was devoted to maintaining indefensible positions. On the other hand, willy-nilly, the controversies themselves stimulated an interest in science and led to accumulation of facts and to the carrying out of valuable experiments, even by those who were to be proved wrong by them.

The practical man: contempt of theory

Technical progress was almost as much hampered by the handicraft traditions of the practical man. Only in stirring times, and then only where the possibilities of immediate profit were obvious, could new ways be easily adopted in preference to old. The story of the manu-facture of iron and steel show how persistently obstructive were the men of practical experience. Up to a point of course they were justified. When a process is not under-stood but is known to work there is always a real danger in any variation, however apparently supported by theory. True, the difficulties and dangers were often avoided with experiment and care, but it was difficult to know in advance that they would be. The British Admi-ralty in their steadfast opposition to the steamship and the iron ship were only expressing a rational fear of novelty in dealing with that very dangerous element—the sea. Over and over again in the nineteenth century such resis-tance was overborne only to find the opinions of the innovators of an earlier generation set hard against the changes advocated by a later one. The steel ship was fought as hard against in 1890 as the iron ship had been in 1850.

More significant perhaps in retarding progress than the vested interest in established techniques, which was as much mental as financial, was the contempt of the practical man for what he called theory, which covered all applications of science. This philistine attitude produced a reaction even among scientists not dissimilar to that which the corresponding attitude to beauty, art, and humanity produced on writers as different as Dickens, Carlyle, Ruskin, and Morris.

Industrial contamination. The ideal of pure science

Scientists and humanists alike turned away from an industry which seemed to have only sordid aims and seemed incapable of realizing the hopes that had been placed on it in the seventeenth and eighteenth centuries. This withdrawal, which became more marked as the Industrial Revolution gained momentum, reinforced the ideal of pure science. This period in which science claimed a virtual immunity from practical concerns is one to which so many scientists today, especially in the older universities, look back with such regret, usually without considering its limitations. They look back in vain, for it is as far beyond recall as are the Middle Ages. The general development both of science and industry leads inevitably to ever more intricate forms of organization; our task is not to resist this tendency but to understand it and to see to it that what organization is necessary is used effectively and for good ends.

Conclusions

Throughout the nineteenth century the scientist had enough to do to establish himself as an acceptable member of an old and tradition-ridden academic society, where he tended to imitate his colleagues in other faculties and to draw aside from the industrial world partly from intellectual snobbery, partly from a more worthy disgust at the unashamed money-hunting and philistinism of the business man. This retreat effectively determined the ethic of what was to be a new profession, comparable with the age-old professions of law and medicine and like them primarily devoted to the service of the upper classes. Indeed the withdrawal of the scientist from contact with industrial realities was derived from the existence of a profound contradiction at the root of nineteenth-century society. It was difficult, indeed almost impossible, to satisfy the absolute need for science to draw its inspiration from practical human enterprise, without engaging in, or at least conniving with, the universal corruption of industrial effort to unworthy ends.[1]

[1] This contradiction, visible only in particular contexts to the scientists, had been seen as part of far wider contradictions of capitalism by Marx and Engels. We find in the *German Ideology* (1844) the following characterization of the new industrialism:

'It [large-scale industry] destroyed as far as possible ideology, religion, morality, etc., and where it could not do this, made them into a palpable lie. It produced world-history for the first time, in so far as it made all civilized nations and every individual member of them dependent for the satisfaction of their wants on the whole world, thus destroying the former natural exclusiveness of separate nations. *It made natural science subservient to capital* and took from the division of labour the last semblance of its natural character. It destroyed natural growth in general, as far as this is possible while labour exists, and resolved all natural relationships into money re-

Conclusions

The contradiction is indeed an inescapable one under the conditions of capitalism and finds its resolution only in a change to a social system where the scientist can work practically with and for the whole people. Throughout the nineteenth century it only served to present the scientist with increasingly painful dilemmas.

The choices they made are revealed in the study of their lives and actions. Only a few great scientists like Carnot, Liebig, Pasteur, and Kelvin managed to contribute directly to the economic progress of their time. Most of these, for one reason or another, themselves escaped the corrupting influence of wealth but all contributed indirectly to the wealth of those capitalists who exploited the ideas. Others like the Siemens brothers threw themselves wholly into the creation of new scientific monopoly industries. Both groups were, however, exceptional: as the century drew on the main body of scientists worked in an increasing divorce from the great industrial developments of their time.

lationships. In the place of natural towns it created the modern, large industrial cities which have sprung up overnight. Wherever it penetrated, it destroyed the crafts and all earlier stages of industry. It completed the victory of the commercial town over the countryside. Its first premise was the automatic system. Its development produced a mass of productive forces, for which private property became just as much a fetter as the guild had been for manufacture and the small, rural workshop for the developing craft. These productive forces received under the system of private property a one-sided development only, and became for the most part destructive forces; moreover, a great multitude of such forces could find no application at all within this system.' (My italics, J.D.B.)

K. Marx and F. Engels, *The German Ideology*, London, 1938, p. 57.

Conclusions

The more far-seeing of these scientists were, nevertheless, contributing their share indirectly in the essential work of systematizing and rationalizing pioneer observation and experiments and thus providing a basis for new advances of industry. Clausius and Gibbs followed Carnot and Mayer and helped to found a new chemical industry. Maxwell and Hertz followed Oersted and Faraday, giving rise to radio and to all of modern electron physics. In spite of this it must be admitted that much of the intellectual effort going into academic science in the nineteenth century wasted itself on sterile exercises on outworn themes. A perusal of old numbers of scientific journals makes this too deadly clear. Even the one justification academic science offered, the contribution to its teaching of new generations of scientists, was often nullified by the dryness and dogmatism of its teaching.

Social consequences of science. Industrialization and imperialism

Enough has been said to indicate the main factors which served to favour or hinder the advance of science and technology in the nineteenth century and the examples given bring it out more clearly than any generalization. It now remains only to discuss the bearing of that advance on the people who lived through that decisive period in human history. The enormously greater understanding of nature, the enormously greater powers given to man had not produced, as their hopeful originators had anticipated, the general well-being and happiness of man.

The transformation was limited to a great multiplication of human beings, particularly in North-Western

Conclusions

Europe and North-Eastern America. The middle classes and even some better-paid workers were living at a level of consumption never reached for such a large body of people at any other time, but, considering the conditions of congested and smoke-laden cities, it is doubtful whether the majority even of these were living at a higher level of health or amenity. For the great mass of new factory and mine workers—men, women, and children; for the temporarily or chronically unemployed, the reserve army of industry, conditions were far worse than for a comparable body of humanity at any previous time. Work was still hard and unremitting. J. S. Mills' dictum 'It is doubtful whether any invention has relieved man of any painful toil' was almost as relevant at the end of the century as when it was made. And for most there was no more security, certainly no more economic security, at the end of the century than at its beginning. The working classes in the industrial countries, the peasants of the backward countries, now coming for the first time under the influence of the world market, received little from the new science and technology. They remained in fact quite outside it.

This is one aspect of nineteenth-century science that is generally overlooked, because until very recently it was taken for granted as a necessary aspect of the science of all time. However much it increased in scope during the century, science remained throughout the preserve of a small minority of people in a very small section of the world. Only the industrial countries of Europe and the newly industrialized parts of America contributed to modern science. The rest of Europe and all of Asia and

171

Africa were left out, though the exploitation of their people was essential to the very existence of industrial capitalism.

Even in the capitalist countries themselves, though the rising manufacturers and engineers won a place for themselves in science, it was still open to a very small section of the population. Samuel Smiles' efforts in *Self Help* to prove the contrary only succeeded in producing pathetic examples of how impossible it was for a working man to be anything but a praiseworthy amateur[1] unless he managed by early success to lift himself entirely out of his class. Actually, as the century wore on, it became more and more difficult for the poor outsider to contribute to invention, let alone to science. George Stephenson was the last of the great workmen-inventors in Britain. Edison, for all his casual and self-supporting youth, cannot be called a working man and he was the last of the self-taught inventors. These national and social limitations must have prevented the contributions to science of all but a very small fraction of those competent to do so. In other words the potential advance of science was many times larger than what was actually achieved.

Interactions of scientific and economic factors

I have traced in this essay something of the relationship of various intellectual and social steps that have marked the advance of science and technology in the nineteenth century. The close interrelation between economic and

[1] See in particular S. Smiles, *Robert Dick, Geologist and Botanist*, London, 1878. He was a gifted geologist who was obliged to live out his whole life as a baker in Thurso.

social factors on one side, and scientific and technical ones on the other does not admit dispute. Which should have the priority is not so easy to see. The traditional view of the great transformation of science and industry of the nineteenth century is that it is a direct consequence of the discoveries and inventions of great men whose activity can be conveniently accounted for by invoking an intangible genius. The converse picture—that the ideas arose entirely out of the operations of economic law—is nowhere seriously maintained, though it still serves the professional anti-Marxists as a convenient Aunt Sally. Marx and Engels themselves were from the outset very far from such a crude analysis. They saw natural science as a part of a wider science of humanity and found in industry the link between natural science and man.[1] They

---

[1] '. . . Natural sciences have developed an enormous activity and appropriated to themselves a steadily increasing field. Philosophy, however, has remained as strange to them as they have remained to philosophy. Their momentary union was only a fantastic illusion. The will was there but the means were lacking. [This refers to the demise of the great German *Naturphilosophie*—J.D.B.] Even the writing of history only gives incidental attention to natural science as an element of enlightenment, as the utility arising from individual great discoveries. But the more science has practically intervened in human life and transformed it through industry, thereby preparing the way for human emancipation, the more it has been obliged to complete a process of dehumanization. Industry is the real historical relation of nature, and therefore of natural science, to man. Hence if natural science is understood as an external revelation of human powers, the human essence of nature or the natural essence of man will be understood and hence natural science will lose its abstract materialist or rather idealistic tendency and will become the basis of human science as it has already become, although transformed

understood, indeed they were the first to analyse, the close reciprocal relation between industry and science.[1] Recognizing that technical advance and the economic and social changes it induced sprang ultimately from science they saw that at the same time the progress of science in character as well as rate depended on the social and economic conditions. These conditions under capitalism were not favourable to a uniformly accelerated growth of science.

The very opportunities given to technical advance, and consequently to science, by the system of private enterprise for profit led to a situation which checked its further application. The profitability of the capital ex-

---

[through industry—J.D.B.] the basis of actual human existence. One basis for life and another for science is *a priori* a lie. . . . In time natural science will include the science of man in the same way as the science of man will include natural science. There will be only one science.'

Marx/Engels Gesamtausgabe, Berlin, 1932; Karl Marx, *Okonomisch-Philosophische Manuskripte*, 1844, pp. 122–3.

[1] Engels alluded to it for instance in a letter fifty years later:

'If, as you say, technique largely depends on the state of science, science depends far more still on the *state* and the *requirements* of technique. If society has a technical need that helps science forward more than ten universities. The whole of hydrostatics (Torricelli, etc.) was called forth by the necessity for regulating the mountain streams of Italy in the sixteenth and seventeenth centuries. We have only known anything reasonable about electricity since its technical applicability was discovered. But unfortunately it has become the custom in Germany to write the history of the sciences as if they had fallen from the skies.'

Marx and Engels, Selected Correspondence, 1943, Letter 229, Engels to H. Starkenburg, 25th Jan. 1894, p. 517.

pended on once new techniques tended to hold up further technical development. Through the chaotic and cruel mechanism of slump and boom the old was periodically shaken off to make way for the new. But because the benefits accrued primarily to a limited class, demand could never be effective enough to secure the rapid and continuous advance that was technically and scientifically possible. Consequently throughout the century the gaps between what was attainable and what was achieved widened rather than narrowed. The most telling proof of this was the way in which, in response to the urgent needs of war, the speed of technical advance could be immensely accelerated; and so, though necessarily more slowly, could that of science, as the developments in nuclear energy and antibiotics in our time have shown.[1]

We cannot indeed expect to find any simple formula in which to comprise the relation of science and technology. All that can be done, and what I hope has to some extent been done in this essay, is to bring out, through examples, the reciprocal nature of the relation, to show both science and technology as human enterprises linked with current economic, political, and cultural forces. We find in this nineteenth century the close association, in place and in personalities, between the social transformation brought about by the capitalist economic system and the development of science. We note that science at the beginning of the century depended for its existence on the new industrial forces, while at the end it had so grown that it could be used to create new industrial

[1] See my essay 'Lessons of the War for Science' in *The Freedom of Necessity*, 1949.

forces. This large-scale scientific industry was in turn to produce economic social strains which were to threaten the system that gave them birth. Twentieth-century science, now integrated with industry, is proving too big for the system of private enterprise that gave birth to it two hundred years ago. Bringing out these connections is not equivalent to saying that science is nothing but economics, as is often hastily assumed; it is simply warning us to look out for the economic causes and economic consequences of scientific advance.

The advance of science. Lessons for our time

The fact that these causes have not been apparent at the time does not make them less real or important; it only means that for lack of knowledge the course of events was out of human control. In analysing the past there is always the danger of projecting present ideas backwards. Though almost any modern scientist could see the limitations of advance of science a hundred years ago, only a very few of the men of that time could do so. Babbages and Pasteurs were rare. The more serious danger, because it operates now, is the opposite one of projecting the past into the present and assuming that the stumbling rate of scientific progress of the last century, produced by a small élite and divorced from the common man, is the only way science can ever advance. Those who believe this, and far more do unthinkingly than have ever examined the question, are helping by their ignorance to hold up the progress of science. What they cannot do is to avoid the economical and social consequences of the technical advances and these for lack of rational control are extremely unpleasant and dangerous, as the experience of our own age has shown.

Even in the nineteenth century the appearance of large-scale industry brought about an increase in social and political strain. On the one hand the working class was becoming increasingly organized and conscious of its wrongs and its strength, on the other the different capitalist powers were showing an increasing rivalry in their search for markets. The forebodings of the wars and revolutions to come could already be felt under the surface of an apparently prosperous civilization.

We are now in the midst of a struggle which is evidently a major turning-point in the history of mankind. In whatever form it may show itself it is fundamentally one to determine whether the new forces that science has shown men how to use will be used in an orderly way for the benefit of all, or will be monopolized and directed as they have been in the past for the benefit of a few. In an era of atom and hydrogen bombs, of bacterial and radioactive poisons, of jet planes and supersonic rockets, the cost of the struggle, if it is allowed to break out into another war, would be a terrible one. But whether it has to be faced or not there can be no doubt as to the final state. In the long run the constructive rather than the destructive use of science is bound to prevail. The human potential available to a society where the whole available population is educated to think and work together scientifically vastly exceeds the power of any élite bent on preserving their position by the elaboration of the most efficient means of destruction.

What does matter is that people should understand in

time the character of the world they live in so as to ensure that the process of transition to a reasonable and constructive world should be rapid yet not catastrophic. One part of this understanding is the appreciation of the place of science and technology in the present world and for that an excursion into its past may be a useful preliminary.

MOLECULAR ASYMMETRY

Chapter VII

MOLECULAR ASYMMETRY[1]

I CONSIDER it a very great honour for me to speak on the occasion of the fiftieth anniversary of the death of Pasteur about the very work with which he began his scientific career. It was indeed more than this, for, as I hope to show, this question of molecular asymmetry not only furnishes the key to all Pasteur's subsequent work, but also marks an important turning-point of the science of the nineteenth and twentieth centuries.

In order to be able to understand clearly what Pasteur achieved and, what is more important, what he has to teach us for the present and the future, it is necessary first to examine the scientific antecedents of Pasteur's work so as to place his chief discovery on molecular enantio-morphism in its contemporary scientific frame. Obviously this must imply a very extensive study. To be able to do it full justice I would have to know far more than I do of the history, not only of chemistry, but of physics and

[1] The numbered footnotes are from my original lecture. Those added for this edition are asterisked. Material written subsequently in the text is enclosed in square brackets [ ].

crystallography of the nineteenth century, and even as far back as before the seventeenth century.

I believe, however, without such a thorough investigation it is still possible to set out the main roots of the family tree, as it might be called, of Pasteur's discovery. In the account of this story as it appears in the biographies of Pasteur* or in general books on the history of science, the main accent is on the brilliance and intuition of the young genius. Enough account is not taken of the fact that Pasteur's discovery was entirely logical and held, from the very start, a central place in the history of science. Pasteur himself was fully aware of this, witness the lectures he gave in 1860 and again in 1883, where he traced in a very comprehensive way the whole history of his discovery.

Accordingly, it is not necessary for me to do more than indicate (the chart) where the main lines leading to Pasteur's discovery can be found. It arose, like all the great discoveries of science, at the meeting-place of hitherto distinct disciplines. Its largest root lay in the chemistry of the tartrates and racemates, drawing its material support from the age-old wine industry. Its immediate origin was, however, physical. Pasteur himself distinguished between the lines of crystallographic investigation stemming from Haüy and those of optics started by Malus' discovery of polarization. Haüy was the first to use the quantitative hypothesis according to which crystals were formed of little crystalline molecules (*molécules intégrantes*)—little parallelepipeds piled together in a regular way.

* R. Vallery-Radot, *The Life of Pasteur*, London, 1920; R. J. Dubos, *Louis Pasteur, Free-lance of Science*, London, 1951.

Molecular Asymmetry

This concept of Haüy's is the key to the understanding of the work of Pasteur because he always understood and grasped the idea of a molecule not as a hypothetical entity, logical and indefinite, but as a real and solid object that would occupy a definite space. Because he had this completely concrete idea, he was able to see further than crystallographers much more learned than himself who were always thinking in terms of purely mathematical and abstract forms. Haüy was also in a more personal way the source of Pasteur's crystallographic ideas. He absorbed them from his teacher Delafosse, himself a pupil and assistant of Haüy.

It was Haüy himself who made the first approach to the problem of enantiomorphism by his discovery in 1810 of the hemihedry of quartz. He found that certain quartz crystals had small facets which did not depend on the primitive forms,* which were arranged like spiral

---

* Haüy noticed that in breaking up Iceland spar crystals, irrespective of their external form, he always obtained the same small rhombohedra. He called these the *primitive forms* out of which any known face could be constituted rather as the sloping sides of the great pyramid could be determined from a stepwise piling of cubic blocks of stone. It is worth quoting his next stage in full for it states clearly the doctrine Pasteur was to make such use of.

'Or, la division du cristal en petits solides a un terme, passé lequel on arriverait à des particules si petites, qu'on ne pourrait plus les diviser sans les analyser, c'est-à-dire sans détruire la nature de la substance. Je m'arrête à ce terme et je donne à ces corpuscules que nous isolerions, si nos organes et nos instruments étaient assez délicats, le nom de *molécules intégrantes*. Il est très probable que ces molécules sont les mêmes qui étaient suspendues dans le fluide où s'est opérée la cristallisation. Au reste, elles seront tout ce qu'on voudra. Toujours

staircases, some crystals bearing them turning to the right, others to the left. These he called plagihedral facets.*
The plagihedry of quartz was the first indication of the existence of substances which crystallize like corkscrews or snails, with a spiral symmetry. Thus, for each structure of this type there is an inverse form identical in all other aspects, in the same way as a left and a right hand or an object and its reflection in a mirror.

During the same period the optical history of Pasteur's discovery began with the discovery of polarized light by Malus in 1808.† This led in turn to the discovery by Arago in 1811 of the rotation of the plane of polarization by quartz. Four years later, Biot found that this property of turning the plane of polarization was shared by certain organic solutions, in particular tartaric acid or its salts.

These two facts, one from crystallography and the other from optics, had one thing in common—quartz. J. F. W. Herschel at Cambridge noticed in 1820 that the

est-il vrai de dire, qu'à l'aide de ces molécules la théorie ramène à des lois simples les différentes métamorphoses des cristaux.'
Observations sur la Physique, 1793, Vol. 43, p. 108.

* The full symmetrical development of crystal faces Haüy referred to as *holohedral*; if half of any form were missing this was referred to as *hemihedry*; if three-quarters, *tetartohedry*; in general, imperfect forms were called *plagihedral*.

† This discovery itself, though apparently accidental—Malus was idly looking at the sunset light on the Tuilleries' windows through a clear piece of Iceland spar—had itself a long history. It involved crystals through another property, their double refraction discovered by Bartholin in 1669, studied by Newton, and explained by Huyghens in 1678.

direction of rotation of the polarized light was related to
the arrangement of Haüy's plagihedral facets. That is to
say, if the facets were on one side of the crystal it turned
the plane of polarization to the right; if they were on the
other side, to the left. Thus he had established the first
link between crystalline forms and the optical rotation.
Herschel saw clearly how important this discovery might
be—at the end of his paper he wrote:

'The fact above recorded is interesting in another point
of view. It may lead us to pay a minuter attention to those
seemingly capricious truncations on the edges and angles
of crystals which appear to be commonly regarded as
the effect of accidental circumstances prevailing during
their formation. It may be so, but the much greater
comparative frequency of some of them than others is
an indication at least of greater facilities afforded to the
decrements by which they are produced, by the constitution of their molecules, and it is not improbable that an
accurate examination of them may afford us evidence of
the operation of forces of which we have at present no
suspicion.'[1]

It fell to Pasteur, twenty-eight years later, to fulfil this
prophecy and throw light on the constitution of molecules from a study of their crystalline forms.

He achieved this by taking as his starting-point a careful
study of a group of salts which had already long proved
of great physical and chemical interest—the tartrates.

Thus it was tartar that provided the third main source
of Pasteur's discovery—the chemical source. This source

[1] *Transactions of the Cambridge Phil. Soc.*, Vol. I, 1822, p. 51.

actually goes back to most distant antiquity. Tartar* was one of the few definite chemical substances recognized by the Greeks and probably before them by the Egyptians and Babylonians; and the study of tartar has always been a fruitful one in the history of science, even up to our own day as I hope to show in what follows. Tartar owed its importance to the fact that it was a by-product of the greatest chemical industry of antiquity and the Middle Ages—the fermentation of grapes. It was therefore available in considerable quantities; and furthermore, as it crystallized easily, it could be obtained in a pure state. Already in the seventeenth century, Seignette had prepared the double salt of potassium and sodium tartrate to which he had given his name and which was well known in medicine.

Scheele, in 1770, starting with this salt, prepared the first organic acid, tartaric acid. The account of this was significantly the first published communication of this brilliant Swedish chemist who must share with Priestley and Lavoisier the discovery of the true theory of combustion and hence rank as one of the founders of modern chemistry. Tartar had, therefore, already played an

* E. Andrews writes in his fascinating *History of Scientific English*, New York, 1947, p. 187:
    'Another example of far-reaching ramifications comes from viniculture. A white incrustation collecting on the bottom of the fermenting vats was called *durd* in Persian. The Arabs corrupted this to *tartar*, and when the substance was purified by the alchemists it was called "tartaric acid". This word is not only cognate with English "dirt", but also with the vulgarism "turd", feces. In view of this fact it is curious that the Romans called the lees of wine *faex*, singular of *faeces*.'

186

important part in the development of chemical ideas in the eighteenth century.

A more immediate effect arose from this discovery of Scheele's; it gave rise to an industry, the manufacture of tartaric acid, which soon found uses in textiles and in medicine. This industry was very simple and widespread at the beginning of the nineteenth century and was found in every wine-growing district. It was from it that there came, in its turn, a new key discovery—that of racemic acid.

I wish to emphasize this point because if there had not been a tartaric acid industry the world would have had to wait far longer for the discoveries of Pasteur and the enormous consequences which follow from them. It was in the tartaric acid factory at Thann, in Alsace, that Mr. Kestner distinguished in 1818 the first crystals of a new acid, which we now call racemic acid, but which was originally called paratartaric, thannic or vosgic acid. Berzelius says of it:

'Someone at Thann, a little town in the Department of the Haut-Rhin, who was employed in the large-scale preparation of tartaric acid, found that together with the ordinary tartaric acid a small proportion of another acid crystallized out, one which was less soluble. This person took it to be oxalic acid and tried to put it on the market as such. John was the first to mention it in 1819 [*Dictionaire de Chimie*, Vol. IV, p. 125]; he pointed out that the substance was neither tartaric nor oxalic. He gave it the name of acid of the Vosges. Gay-Lussac, who visited Thann in 1826 and was given a certain amount of this acid by the manufacturer, made some investigations which led him to the conclusion that this was not tartaric acid although its saturation capacity only differed by a few thousandths from that of tartaric acid.'[1]

This acid, a new organic acid the fame of which quickly spread, was chosen by the most eminent chemists of the period as a good example on which to try the new possibilities of organic analysis. It was Gay-Lussac who was the first to see that this was a hitherto unknown acid, strongly resembling tartaric acid, but differing from it in a few respects, notably in the solubility of its salts.

About 1830 Berzelius himself began the chemical study of the salts of tartaric and racemic acids. He found that they were identical from the chemical point of view, and in this identity he found decisive support for his great discovery of chemical *isomerism*: namely that two different chemical substances could exist, each possessing the same elements in the same proportions:

'This research shows therefore that racemic acid not only has the same atomic weight as tartaric acid but also the same composition. It provides us with a new example of an unexpected phenomenon that there are substances which are composed of the same number of simple atoms, but which possess nevertheless different properties.

'The more the existence of such substances is confirmed, the more important it becomes to learn how to recognize their different properties and the form of their crystalline compounds.'[2]

[1] *Annales de Chimie et de Physique*, Vol. XLV, 1830, p. 128.
[2] *Ibid.*, p. 131.

Berzelius had a collaborator, the chemist and crystallographer, Mitscherlich, who was to play a large part in
the story. Mitscherlich had already in 1818 discovered a
phenomenon of great importance to crystal chemistry:
isomorphism—the fact that two substances could have the
same crystalline form but different chemical composition.
Thus it was quite natural for Berzelius to interest his
friend in the question of tartrates and racemates, and for
them to work together on it from the start. It is a very
curious fact, however, that it was not until 1844 that the
results of this work were published. Possibly this was
because they had not arrived at a satisfactory conclusion.
But according to the correspondence between Berzelius
and Mitscherlich, it is evident that already in 1830
Mitscherlich had measured the crystals of tartrates and
racemates and had made the puzzling but apparently
decisive observation that the double tartrate of potassium
and sodium was isomorphous with the corresponding
racemate of potassium and sodium. In May 1830 Berzelius
wrote to Mitscherlich asking him to crystallize the double
salts of potassium and sodium of tartaric and racemic
acids and to examine their forms. He ended with the
words:

'Should the form be different, then the awkward
difficulty (*der schwere Knoten*) of the dual relationship of
bodies with the same composition would be solved in
a simple and possibly correct manner.'[1]

He hoped thus to find the crystallographic differences
between isomeric substances and recognized that in

[1] *Collected Writings of Eilhard Mitscherlich*, 1896.

succeeding he would have resolved the difficult problem of isomerism. But he did not succeed in doing it; a year later he wrote:

'No tartaric acid salt resembles the other corresponding Vosgic (racemic) acid salt with the exception of the sodium ammonium salt of tartaric and vosgic acid; both double salts have the same form, i.e., that of Seignette salt; I originally believed that I had found in this salt a means of converting Vosgic acid into tartaric acid; however when I transferred the Vosgic acid of this sodium ammonium salt to other bases I again obtained Vosgic acid salt. From this it is evident that we are not yet clear [*nicht im Reinen*] on this point. I urgently beg you to continue your researches in this direction; for finally you will probably find the true cause of the differences between these salts.'[1]

However, neither Berzelius nor Mitscherlich were ever able to distinguish between the crystals of the racemates and those of the tartrates.

Thus, already in 1830, the importance of this problem of the tartrates was well recognized by the leading scientists of the period. But they had not solved it and eighteen years passed without their having found the secret.

It was Pasteur who was to find it by making the final and decisive discovery.

Mitscherlich had later tried to see whether this apparent identity between the double salts of the tartrates and the racemates extended to the property of rotation of the

[1] *Ibid.*, p. 96.

plane of polarized light; but he found the contrary—
the tartrate was active to the right, the racemate inactive.
He communicated this result to Biot and asked him to
confirm it. Biot did this and communicated the result to
the Academie in 1844. He expresses himself in these terms
in a note which Pasteur found in 1846 when he was at
the École Normale, and which made a strong impression
on him:

'It is known that tartaric acid possesses the power of
rotation and imparts the property to all its saline combinations. Paratartaric (racemic) acid, on the contrary, although having the same composition by weight, does
not possess this property and does not impart it to any
of its saline combinations.

'M. Mitscherlich decided to examine whether this
difference would still exist in circumstances when the
two substances compared were similar, not only in
chemical composition, but also in their crystalline form
and in their physical properties. He found these conditions
united, with remarkable identity, in the double salts
formed by these two acids with sodium and with ammonium. The results he obtained from these two substances
are described by him in the following note:

'"The paratartrate and the double tartrate of sodium
and of ammonium have the same chemical composition,
the same crystalline form with the same angles, the same
specific weights, the same double refraction and, as a
consequence, the same angles between optical axes.
When dissolved in water, their refraction is the same.
But the tartrate in solution rotates the plane of polarized

light and the paratartrate does nothing as M. Biot found
for the whole series of these two types of salt; but here
the nature and number of the atoms, their arrangement
and distances are the same for the two substances compared."'

This was in direct contradiction to the idea that the
crystalline form is an indication of the molecular constitution.

For Biot, this only meant that the effect that he had
discovered, optical rotation, was the only way of dis-
closing molecular structure. Later he commented, and
Pasteur certainly read this commentary, that:

'The physical properties resulting from the building-
up of these masses such as sound vibrations, crystalline
forms and double refraction, give no information on
the character of molecules or at least they have not been
deduced up to the present mechanically from them. The
only phenomena whose observation and measurement
might be legitimately related to the constituent molecular
groups themselves, thus seem to me in the present state
of our knowledge, to consist uniquely in the deviations
that a large number of substances, all in fact of organic
origin, impose on the plane of polarization of light rays,
independently of their random state of aggregation.'

Pasteur never saw things in this way because he was a
crystallographer of the school of Haüy and he did not
wish to admit that the faces shown by crystals were not
an expression of the molecular structure.

He was already convinced of this idea when he was at

the École Normale and he started his first work with the conscious aim of clarifying these contradictions which to him could only be apparent. In a lecture which he delivered thirty-five years later, in 1883, he said:

'Scarcely had I left the École Normale than I planned to prepare a long series of crystals in order to determine their forms. I considered tartaric acid and its salts as well as those of paratartaric acid for the two reasons that the crystals of all these substances are as beautiful as they are easy to prepare, and, on the other hand, I could at each instant control the accuracy of my determinations by referring to a memoir of an able and very precise physicist, M. de la Provostaye, who had published an extensive crystallographic study on tartaric and paratartaric acid and their salts.'[1]

It is strange to note that this scientist had done very extensive work on the tartrates without finding what Pasteur discovered—hemihedry. Perhaps this was defective observation. I believe, nevertheless, that it was due more to the fact that la Provostaye had no preconceived ideas and did not sense the great importance of this work; it is this that constitutes the genius of Pasteur. Pasteur had two principal ideas already fixed in his mind when he was at the École Normale. He was, as a chemist, in complete agreement with Chevreul's definition of chemical species, essentially a very simple one: 'Amongst compounds, the species is a collection of objects (*êtres*) identical in composition, proportion and arrangement

[1] 'La Dissymetrie moléculaire', lecture delivered before the Société chimique de Paris, 22nd December 1883. *Œuvres*, Vol. I, pp. 369–80.

of the elements.' This was Pasteur's idea of the structure of molecules. On the other hand, he took from Haüy the idea that the crystal gives a picture of this arrangement. From the very outset of his scientific career, Pasteur foresaw clearly two methods of approaching the study of substances: the *chemical* method and the *physical* method. In the notes for his unpublished book on crystallography can be found:

'In order to approach the goal towards which so many efforts are striving, there are two paths to be followed. One can start from the chemical properties, properly speaking, alter the substance by various reagents, carefully study the resulting products, and then try to infer the dispositions of the atoms from the way in which their original arrangement dissociates. This analytical progress is powerfully aided by synthesis, that is the study of the processes necessary to reconstitute the arrangement in question starting from more simple ones. This is what can be called the *chemical method* proper, an extremely extensive method which is that almost universally followed by chemists. But there is another way of tackling the problem. It consists above all of not altering the substance and of investigating scrupulously its properties as an intact unit (*interroger scrupuleusement ses proprietes toutes faites, si je puis ainsi parler*), notably those which are most directly dependent on the mode of its internal arrangement. And as certain physical characteristics, such as the crystalline form, or the effects of changes imposed on light when it passes through crystalline substances or solutions, must play the biggest part, this

method of proceeding can be called the *physical method.*
Even though it is more restricted than the chemical
method, it is perhaps more precise and more certain.
From inclination and no doubt also by the fortune of
circumstance, it is this latter method which I have more
specially followed in my researches, while still not
neglecting the former.'[1]

These ideas of Pasteur's have lost nothing of their
power of inspiration during the course of a century; the
two methods, chemical and physical, have merely come
closer to each other.

Let us now examine how Pasteur, at the start of his
scientific career, set out to attack the problem of the tartrates which so many great scientists had attempted and
failed to solve.

For this purpose we now can use, besides the illuminating and well-known lectures given by Pasteur himself
in 1860 and 1883, the even more valuable pages of the
very notebook—Pasteur's first research notebook—in
which he wrote from day to day his programme, his
ideas, and his observations as he made them.[2]

At the head of the first page (Fig. 6) can be seen:

Tartrates (Questions to be solved)

There follows the formulae of the salts which he was

[1] 'Pages inédites de Pasteur.' *Œuvres*, Vol. I, pp. 392–3.

[2] These notebooks, fortunately preserved for almost a century,
have remained unpublished. I owe the honour of having examined
them to the kindness of M. Wyart, Professor of Mineralogy at the
Sorbonne. I saw them only a day before this lecture; they deserve a
detailed and thorough study.

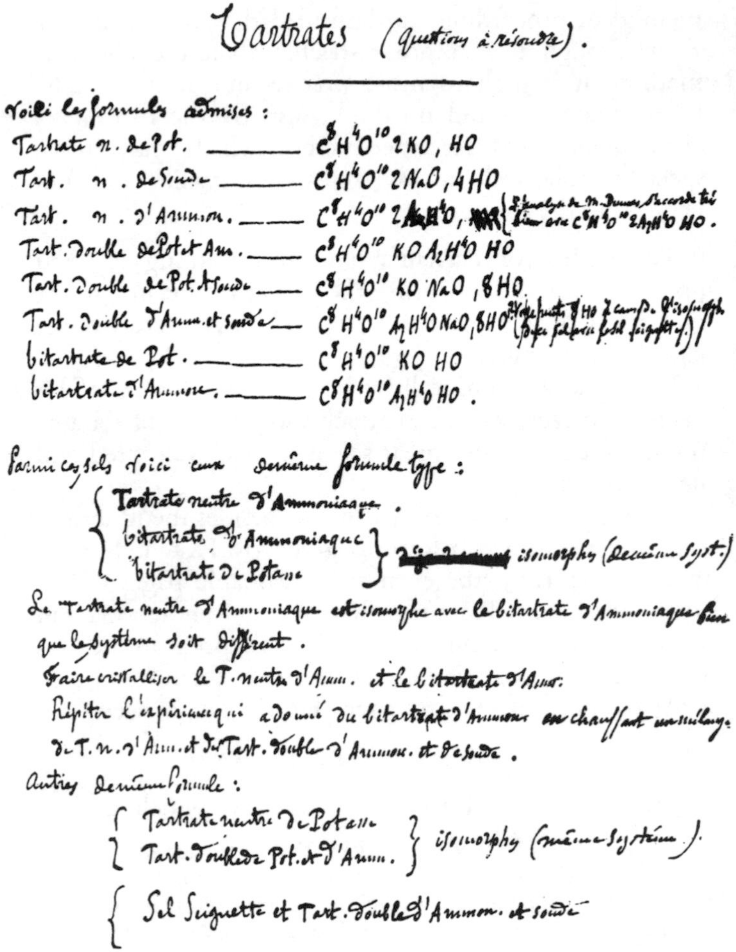

Tartrates (questions à résoudre).

Voici les formules admises :

Tartrate n. de Pot. _____ $C^8H^4O^{10}$ 2KO, HO

Tart. n. de Soude _____ $C^8H^4O^{10}$ 2NaO, 4HO

Tart. n. d'Ammon. _____ $C^8H^4O^{10}$ 2AzH⁴O, ... { analyse de M. Dumas s'accorde très bien avec $C^8H^4O^{10}$ 2AzH⁴O HO.

Tart. double de Pot et Am. _____ $C^8H^4O^{10}$ KO AzH⁴O HO

Tart. double de Pot et Soude _____ $C^8H^4O^{10}$ KO NaO, 8HO

Tart. double d'Amm. et soude _____ $C^8H^4O^{10}$ AzH⁴O NaO, 8HO ...

bitartrate de Pot. _____ $C^8H^4O^{10}$ KO HO

bitartrate d'Ammon. _____ $C^8H^4O^{10}$ AzH⁴O HO.

Parmi ces sels voici leur dernière formule type :

{ Tartrate neutre d'Ammoniaque .
{ bitartrate d'Ammoniaque } ... isomorphes (deuxième syst.)
{ bitartrate de Potasse

Le Tartrate neutre d'Ammoniaque est isomorphe avec le bitartrate d'Ammoniaque bien que le système soit différent.

Faire cristalliser le T. neutre d'Amm. et le bitartrate d'Amm.

Répéter l'expérience qui a donné du bitartrate d'Ammon. en chauffant un mélange de T. n. d'Amm. et de Tart. double d'Ammon. et de soude.

Autres dernière formule :

{ Tartrate neutre de Potasse
{ Tart. double de Pot. et d'Amm. } isomorphes (même système).

{ Sel Seignette et Tart. double d'Ammon. et soude

FIG. 6. First page of Pasteur's research notebooks on tartrates.

proposing to examine and everything he could find out about their chemical analysis. It can be seen therefore that in his first steps in research Pasteur set to work with defined aims and clear ideas. He soon ran up against difficulties. On a following page we read:

'Among the formulae quoted above that of ammonium sodium tartrate is not given anywhere. I have considered it as the same as that of Seignette salt because of the complete isomorphism between the two salts.

'On this matter I noticed that M. Mitscherlich has said that the double paratartrate of sodium and ammonium and the double tartrate of these bases were completely isomorphous and that one however did not deviate (rotate) the plane of polarization. Now 4 of the same formula M. Gerhardt gives 4 HO for the quantity of water of the paratartrate. Was this the quantity of water in the double T? But then how can its isomorphism with Seignette salt be explained? There is something there to be gone over again, prepare the double paratartrate of S. and Am. Its form and analysis? *Idem.* To analyse the T g[auche] of its bases.

'Ask M. Biot if he still has a specimen of double paratartrate which M. Mitscherlich had sent him and see whether it is isomorphous with the corresponding T or the Seignette salt.'

There follows a minute examination of the crystalline forms of the tartrates. Here is to be found a completely unexpected observation:

'Notes on the n Tartrates of Pot. and Am.

'1. They are both hemihedral.

197

'2. They are isomorphous or at least the two forms though somewhat different are compatible.

'3. They both possess an easy cleavage in the same direction.

'However M. Dumas does not attribute the same formula to them. The analysis of these salts must be undertaken again, first the neutral tartrate of potassium by organic analysis.

'Note on the neutral Tartrate of Pot.,

'M. Hankel (3 anno 1843) Berzelius' annual report p. 135, observed that the n T of Pot is hemihedral and shows electrical polarity on raising the temperature. On raising the temperature one of the ends is positive and becomes negative when the salt cools. The n T of Am is also hemihedral at ends corresponding to the n T of Pot. One batch has shown very beautiful crystals of the n T of Am where this hemihedry is very evident. Tried whether it showed electrical polarity.'

Notice that Pasteur grasped a leading crystallographic fact, hemihedry, but also that he attributes the observation of this property, linked with that of pyroelectricity, to Hankel, to whose works his attention had been drawn by a mere phrase or two in Berzelius' annual reports. He had indeed studied his sources well and appreciated their importance for he then copied whole pages of Hankel into his notebook.[*] What is interesting is that

---

[*] It is an odd fact that in the excitement of the subsequent discovery Pasteur seems to have overlooked entirely his debt to Hankel. Neither in his papers nor in his subsequent lectures on the enantiomorphism of the tartrates does he make any reference to him. Indeed

Hankel himself had not grasped the importance of his own observations on tartar (already in 1840 he had established the pyroelectricity of Seignette salt, *Pogg. Ann.*, Vol. 49, p. 502). But his principal interest was physical and mineralogical. The memoir which Pasteur studied was entitled 'On Topaz' and it was he alone who grasped the significance both chemical and crystallographic of Hankel's observation on pyroelectricity.[1]*

in most histories of science the discovery of the hemihedrism of the tartrates is credited to Pasteur. What is even odder is that Hankel himself, who lived till 1899, seems never to have claimed priority.

Even if his contribution had achieved notice we would still be inclined to attribute the discovery of hemihedrism to Pasteur independently if we had not the irrefutable evidence of the notebooks. This serves to show the importance of preserving original and contemporaneous notes of the progress of science if we wish to follow the *real* steps which have led to great discoveries. Published papers may omit important steps and the memory of men of science, even the greatest, is sadly fallible.

[1] What is more strange is that Berzelius himself, whose interest in this question we have already indicated, did not grasp the importance of Hankel's observation. He wrote in the note to which Pasteur made reference:

'Hankel has undertaken some very interesting researches on the electrical state taken up by various crystalline minerals on changing their temperature. They are shown in Zinc silicate, Axinite, Prehnite, Mesotype, Tourmaline, Topaz, Rutile and Rock crystal as well as in the crystals of tartaric acid and Potassium Sodium Tartrate. However, as the results of these researches belong more to Electrical theory and to physics than to Mineralogy properly speaking, I will allow myself to pass over them.'

Jacob Berzelius, *Jahres-Bericht über die Fortschritte der Chemie und Mineralogie*, 1842, p. 165.

* This is another of many instances in scientific history in which a

However that may be, the pages which follow in the notebook contain admirably detailed studies on the crystallography of tartrates with a particular interest in the hemihedrism. Finally we arrive at the critical case of Seignette salt, the double tartrate of Ammonia and Soda. Pasteur recognized its hemihedry clearly, which is nevertheless easy to miss in this case. We can see with what precision he made his measurements even on the suspicion that the form might be monoclinic (which in fact it is according to the X-ray research of the last few years).

At last comes the critical page, an original and simple document of a great discovery which cannot be read without emotion. It would be well to know the exact day and time when these experiments were made and noted. It was the summer of 1848, an historic year for other reasons, during which events took place which did not leave the young Pasteur unconcerned in his laboratory. In those June days he was already a member of the Garde Nationale.

Let us pass on to the very page (Fig. 7), whose text I have translated literally in its entirety:

great discovery has been missed by a hair's breadth. Berzelius had himself lived through the period which marked the break-up of the one unified Natural Philosophy into a set of specialized sciences. His respect in his old age for these sciences and his feeling of incompetence in those other than chemistry and mineralogy, blinded him to the fact that he held in his hand the clue to the problem which had so passionately interested him in 1830. If he had communicated this to Mitscherlich he might well have anticipated Pasteur. Still he had the scientific scrupulousness to note the facts and thus preserve them in a form in which they could strike the eyes of one who could, as a physicist as much as a chemist, grasp their full significance.

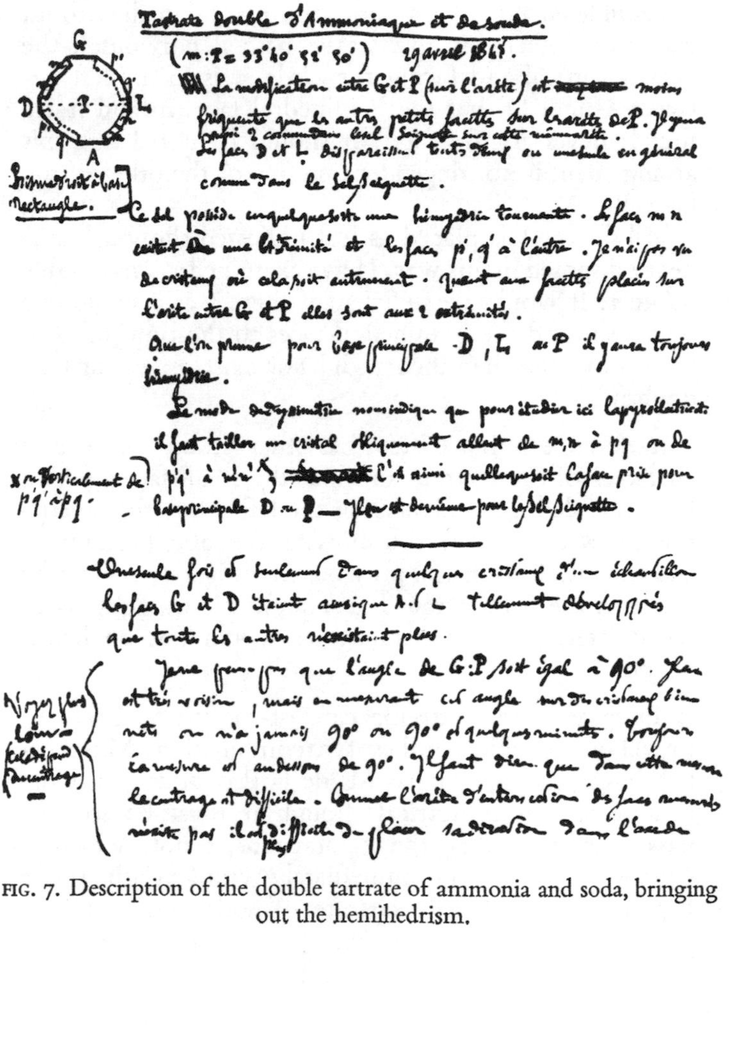

FIG. 7. Description of the double tartrate of ammonia and soda, bringing out the hemihedrism.

'*Double Tartrate of Ammonia and Soda.*

'I have examined a very great number of crystals of the double tartrate of ammon. and soda and in all without exception, when one places the base P horizontal, the face G vertical and facing forwards, it is to the left between G and D that the hemihedral faces m and n are found: never to the right. It should be noted that this arrangement is not dependent on one or the other of the bases P.

'All the crystals placed as just indicated show a hemihedry disposed in this way. I have examined a considerable number. It would be sufficient if there was a single one which showed the hemihedral faces to the right of G to L to prove that hemihedry did not exist or was at least doubtful.'

It can be seen that Pasteur has observed the hemihedral character of the tartrate and that the small facets which have revealed it are always on the left. Note that everything is written out in full showing the same precautions and exactitude as in the previous investigations. Next he passes to the paratartrate (racemate). He thought, as he stated later, that the paratartrate would not be hemihedral:

'In spite of the extreme care of his study, I said to myself on the subject of these two combinations, Mitscherlich has not seen, nor has M. de la Provostaye, that the tartrate was asymmetrical though it must be so; nor have they seen that the paratartrate is not, which is equally very probable. Immediately with feverish ardour I prepared the double tartrate of soda and ammonia and the corresponding paratartrate and I set about comparing their crystallographic forms, with the preconceived idea that I was going to find asymmetry in the form of the tartrate and an absence of asymmetry in that of the paratartrate. Then, thought I, everything will be explained; Mitscherlich's note will have no more mystery, the asymmetry of the form of the tartrate will correspond to its optical asymmetry; the absence of asymmetry in the form of the paratartrate will correspond to the inactivity of this salt on the plane of polarized light: its optical indifference. In fact, I saw that the tartrate of soda and ammonia carried the little faces that betrayed its asymmetry; but when I passed to the examination of the crystal form of the paratartrate my heart lost a beat (*j'eus un instant un serrement de cœur*); all the crystals bore the facets of asymmetry.'[1]

But—and now look at the text of the notebook (Fig. 8):

'*Double Paratartrate of Soda and of Ammon.*

'The crystals are often hemihedral to the left, often to the right,[2] that is the difference between the two salts (*c'est là qu'est la différence des deux sels*).'

The first great discovery has been made. He acknowledges it in half a line: 'There lies the difference between the two salts.'

---

[1] 'La Dissymétrie moléculaire', *op. cit.*, pp. 370–1.
[2] Following this a line is scratched out. It can just be deciphered in the original: 'and very (?) often all the faces are repeated as the symmetry requires' ('*et assez* (?) *souvent toutes les faces se répètent comme le vent la symétrie*'). This shows how he at first went on the wrong track and subsequently corrected himself.

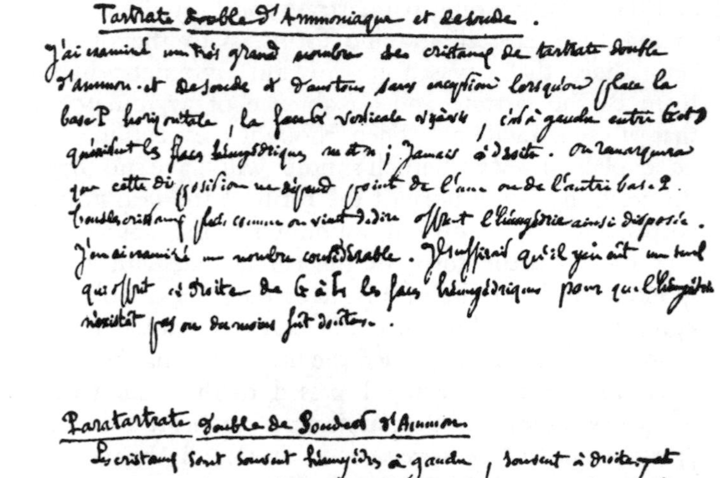

Tartrate double d'Ammoniaque et de soude.

[handwritten paragraph]

Paratartrate double de Soude et d'Ammon.

[handwritten paragraph]

FIG. 8. Decisive page recording the left and right deviation of the plane of polarization of left and right hemihedral crystals separated by Pasteur.

That is not all. The crystallographic problem is solved —the optical problem remains. Look at what follows:

'8 gm. paratartrate (hemihedral on the right) dissolved in 99·9 cm.³ of water in a tube of 20 cm. a deviation

2·8 = 6° 42′ at 17°, to the left ↖ .

'8 gm. tartrate of S. and Amm. in the same circumstances

have given me 7° 54′ at 17°, to the right ↗ .'

He separates and dissolves 8 gm. weight of right-hemihedral crystals, a labour which in itself demands exceptional patience and judgement. He examines them in the polaroscope and observes a deviation. He measures it and then marks almost as an afterthought—under the

line—'to the left (*à gauche* ↖)', two words and a sign.

For the first time in the history of the world the two enantiomorphic forms of a compound have been recognized. This was the second great discovery in a day.

There follows the well-known story of Pasteur and the astonished laboratory assistant to whom he tries to communicate his uncontrollable feelings. But the note-book now returns to its precise and critical manner:

'It is to be noted that it is difficult in spite of the choice of crystals, to separate exactly all those hemihedral to the right from those hemihedral to the left. The hemi-hedry, very evident in the somewhat large crystals, is diffi-cult to determine on the little ones implanted on the big ones. Probably the deviation would be the same for very well-chosen crystals.'

He continues with the tartrate crystals, notes that the rotation to the right is greater and explains it by the difficulty—truly enormous—of effecting a complete separation.

That is all—the page is complete—molecular asymmetry is established.

There are discoveries such as those of Oersted on the relation between electricity and magnetism or even that of Röntgen on X-rays which, once made and published, pass from the hands of those who discovered them to the laboratories of the whole world. That of Pasteur, although its importance was recognized from the outset, was not of this kind. This was because Pasteur alone had the crystallographic and chemical knowledge to develop it and to draw the inferences from it. He devoted himself to this task for the next eight years until he was made a professor at Lille. These years are marked by accurate and at the same time brilliant work. Guided by his ideas on symmetry, he extended and generalized the conception of racemic compounds. He found a new method of separating their enantiomorphic components, that of crystallizing their salts with an active acid or a base, thus giving the two forms different physical properties. It was following this trail of resolution of optically active compounds that he hit on another and purely biological method of separation. This was to turn him away completely from the career of combined physicist, chemist, and crystallographer in which he had already made himself the greatest adept, only to embark on another, that of bacteriology with results which eclipsed all his previous work, which in itself would have been sufficient to establish his immortality in science.

In effect, in 1857, Pasteur found almost by accident that, in his own words:

'I have cultivated little grains (spores) of *penicillium glaucum*, that mould that is found everywhere, on a surface of ashes and paratartaric acid and I saw *l*-tartaric acid appear. Here again simple asymmetry is obtained with an inactive substance; but always again, to arrive at the result it is necessary, you see, to call on the intervention of an asymmetric activity, the asymmetry of the immediate natural products that make up the grains of the mould. Once again these experiments bear witness to the profound line of demarcation between the mineral and organic worlds, since to imitate some fact in nature, that is to prepare a right-handed or left-handed compound, we are constrained to have recourse to a quite special type of intervention, the action of asymmetry. The line of demarcation of which we are speaking is not a question of pure chemistry and of the obtaining of this or that product, it is a question of forces, life is dominated by asymmetric actions of which we feel the enveloping and cosmic influence. Life is the germ and the germ is life. Now who could say what would be the *future* of germs if we could replace in these germs the immediate principles—albumen, cellulose, etc., etc.—by their inverse asymmetric principles? The solution would consist on one side in the discovery of spontaneous generation if such a thing should be in our power, on the other side of the formation of asymmetrical products by the help of the elements—carbon, hydrogen, nitrogen, sulphur, phosphorus—if in their movements these simple bodies could be dominated, at the moment of combination, by asymmetric forces.'[1]

It can be seen from this quotation how Pasteur, while thinking about the phenomena of life and fermentation, linked them with the idea of molecular symmetry, and how his later work followed his thought as a crystallographer at the same time as it served the needs first of industry, and then of the health of beast and man. For Pasteur was one of those great scientists who can make of their work a unity which is at the same time useful and logical. Thanks to Pasteur bacteriology can be said to be born out of chemistry by crystallography.

It is not the aim of the lecture to follow further the work of Pasteur himself. That has been done in many other lectures of this glorious fiftieth anniversary. For me, it only remains to indicate the numerous effects of the utmost importance which the discovery of molecular asymmetry has had on the physical and chemical sciences. As all great discoveries derive from multiple roots, so they also give rise to multiple branches, the most significant of which are indicated in schematic form (the Chart). Let us examine a few of these branches of which the most important is that of chemistry. What is very strange is that the effect on chemistry of Pasteur's work, although altogether revolutionary, was very slow

---

[1] *Œuvres*, Vol. I, p. 377.

to show itself. Pasteur made the principal discovery of molecular asymmetry in 1848, and the first real application of his ideas did not take place until 1874 with the beautiful independent discovery of Van't Hoff and Le Bel of the asymmetric carbon atom.

It can be asked, and it is a very interesting question for the chemical historians, why it took so long to deduce from Pasteur's experiments the consequences which seem so obvious now. I have given much consideration to this question and have read widely in contemporary literature in order to find a convincing answer to it. This leads me to believe that the major reason for this delay can be found in the confused and mistaken state of chemistry with regard to molecular weights and valency.

Pasteur himself wrote the formula for tartaric acid $C_8H_4O_{10}.2HO$ with the old atomic weights, which were 8 for oxygen and 6 for carbon; he could therefore have no idea of the molecular composition of its salts. As a result it was first necessary to wait for chemistry itself to be reformed and for the molecular weights to be understood in the way indicated by Cannizzaro in 1858 on the basis of Avogadro's hypothesis of equal numbers of molecules in equal volumes. After this a long a time was needed for these ideas to be able to penetrate chemistry and to make possible the theory of valency which Kekulé enunciated for the first time in 1859.

But although Kekulé spoke of the tetrahedral carbon atom he did not associate it at all with molecular asymmetry or with Pasteur's effects. This is because the chemists of the nineteenth century, with the exception of Pasteur, thought in terms of two dimensions, not in three—on paper and not in space. Kekulé's model of the carbon atom was a sausage which he denoted thus:

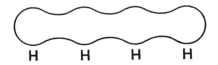

with a hydrogen atom for each slice. But facts incompatible with this view were accumulating. It was probably the very detailed studies by Wislicenius of lactic acid which stimulated Van't Hoff and Le Bel independently in their hypothesis of the asymmetric carbon atom—a completely Pasteurian idea.

From this has followed all the development of the last quarter of a century in organic chemistry: the present-day structural formulae, the formulae derived by the chemical approach in Pasteur's sense (see pp. 194 f.).

Let us now pass on to another consequence of his discovery and take a glance at the development of crystallography.

When Pasteur was at the École Normale, Bravais, mathematician and crystallographer, was at the École Polytechnique. He grasped the importance of Pasteur's discovery for mathematical crystallography. If a crystal possesses at the same time the regular packing of Haüy and the molecular asymmetry of Pasteur, how many ways are there of arranging their relative positions? It was already known, according to Hessel (1830), that crystals could present *thirty-two* symmetrical arrangements of faces. Now the symmetry depends on the number and the arrangement of the axes of rotation and on the planes of reflection. Bravais showed that this symmetry depended on *fourteen* regular space-lattices and thus laid the basis of modern mathematical crystallography.

Soncke developed this idea and added the operations of space symmetry using Pasteur's idea that the whole crystalline edifice can be considered as developed from an arrangement of molecules simply wound round an axis or combined with a translatory movement and therefore forming a helix (spiral) around the axis. He showed that this could be done in *sixty-five* different ways. Two enantiomorphic forms, such as the two types of quartz, right-handed and left-handed, belong to two distinct groups of Soncke.

This work of Soncke leads on in one direction, that of crystal physics, to the work of Pierre Curie. He had started in crystallography, as had Pasteur, by considering the logical consequences of molecular and crystalline symmetry, and he deduced from this the existence of asymmetric properties of crystalline materials in bulk and thus discovered the piezoelectricity★ of quartz.

Thus once again we find quartz, which, like tartar, is one of these protean substances that on closer examination exhibit one new phenomenon after another. The piezoelectricity of quartz has played, thanks to the work of Langevin, a tremendous part in modern electronics, as a distant but absolutely logical consequence of Pasteur's discovery.

Following the other line of purely mathematical

★ See Pierre et Jacques Curie, *Comptes Rendus*, 1880, Vol. XCI, p. 294; *Journal de Physique*, 1894, Vol. III, p. 393.

crystallography we arrive at the solution of the completely general problem of the regular arrangement of atoms in space which Soncke had been unable to find. His groups had axes of symmetry alone, they were all enantiomorphous. To complete the picture it was necessary to introduce planes and centres of symmetry as well, like those possessed by racemic crystals. The different ways of doing this were found simultaneously by the great Russian scientist, Fedorov, by the German, Schoenflies, and by the English, Barlow, who all three found between 1885 and 1894 that this can be done in neither more nor less than *two hundred and thirty* ways; these are the two hundred and thirty space-groups.

It was the combination of this idea of a crystal lattice on the one side and of the physics of radiation on the other that gave rise to another great advance. This was the discovery of the diffraction of X-rays by crystal lattices by Von Laue in 1912 and which the Braggs, father and son, utilized so quickly to work out the atomic structures of crystals. This discovery had far greater immediate consequences than that of Pasteur but it can be seen how in some measure it is the consequence of it. From the moment this discovery was made, all the chemical methods of arriving at molecular structure could be supplemented and ultimately replaced by physical methods. This is because the *physical* method of X-rays— Pasteur's method—is a direct method which considers substances as they are and not as a result of their transformations into other substances. Nearly all the discoveries of modern physics are connected in one way or another with the use of X-rays: the discovery of the electron, the nuclear atom model, the verification of the quantum theory and its application to chemistry. From this stage onwards, the distinction between physical methods and chemical methods disappears into a synthesis which embraces both.

It is by means of the X-ray analysis of crystals we have finally succeeded in determining the precise structure of molecules and thus in finding the exact metrical basis of molecular symmetry. Since Bragg's first work on the diamond and on naphthalene, his successors have striven to penetrate more and more into the complexities of organic structure. A few examples will suffice to show the stages. The first organic analyses, such as those of Mrs. Lonsdale, determined crystal structure on the bases of facts drawn from organic chemistry, which they could only confirm in giving greater precision. In 1935 J. M. Robertson determined the structure of phthalocyanine, a known structure to be sure, but he did it without using a single chemical fact. It is a flat and symmetrical molecule. In 1942, Carlisle and Crowfoot did the same thing for the asymmetrical molecule of cholesterol and thus determined independently of all chemical work its stereochemical constitution. Even the positions of the atoms in Pasteur's classical salt, the double tartrate of soda and ammonia, have been found by Beevers, who has thus confirmed by a physical method the asymmetric character of the tartaric molecule which Pasteur had predicted.*

All this work has verified and made more exact our

* In 1949, since this lecture was given, Parry in my laboratory determined the structure of racemic acid—the origin of the whole dispute which gave rise to Pasteur's discovery.

knowledge of the molecules by giving the distances between the atoms and the angle between the valency bonds. But we have had to wait till recent times—during the war—for crystallography to be employed in finding not only the structure but the very composition of a molecule in advance of chemical analysis. The work of Crowfoot and of Bunn on penicillin—itself arising from those same moulds so dear to Pasteur—marks the triumph of the crystallographic method and justifies Pasteur's premonitions that the crystalline form carries within itself the secret of the molecular structure.

Modern studies on crystal physics show still more the fruitfulness of Pasteurian ideas. The salt of Seignette, of Pasteur and of Pierre Curie, has further very strange and instructive properties. Wul and others have discovered the ferroelectric effect in it, the existence in the 'a' crystallographic direction of an enormous dielectric polarization. This effect, already of considerable use in electronics, has been explained by the X-ray work of Ubbelohde. He has shown that within the range of temperature where this effect is noticed, between the points known as Curie points, the crystal has a lower symmetry, monoclinic—as Pasteur had already suspected—and that it is from this loss of symmetry that the new effect arises. [Finally as a crowning example of the use of the physical method Bijvoet* in 1951 has achieved the apparently impossible task of determining the particular arrangement that corresponds to the normal d. tartaric acid. The difficulty arises because the common use of X-rays does not distinguish between a molecule and its inverse form, it adds

* *Nature*, Vol. CLXVIII, p. 271; *Proc. Roy. Acad. Amsterdam*, Vol. LIV, p. 16.

Molecular Asymmetry

a centre of symmetry to every structure. Bijvoet overcame it by using X-rays of a wavelength almost equal to the characteristic fluorescent rays of an element in the crystal, thus introducing an asymmetric phase lag. The

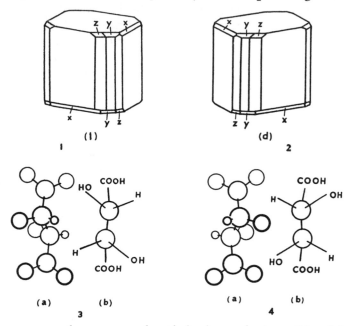

FIG. 9. 1 and 2 are somewhat idealized reproductions of l and d Seignette salt crystals, sodium potassium tartrate hydrate. These were the forms of crystals which Pasteur separated from his crystallization of the racemate. The only difference is in the position of the minute hemihedral faces marked x, y, and z in the figure.

3 and 4 (a) represent the actual configuration of the tartaric acid ions in the corresponding salts as determined by Bijvoet. It will be seen that they, like the crystals, are minor images. 3 and 4 (b) are the conventional designation of the ions as given in chemical textbooks.

differences were small but unmistakable and enabled him to determine that d. tartaric acid had the configuration arbitrarily assigned to it by the chemists, Fig. 9 d and not its inverse Fig. 9 l.]

It would be completely impossible in the course of this lecture to follow further the many branches of the ideas arising out of the fruitful discovery of Pasteur, but I would like to conclude by mentioning one which would have pleased him greatly had he been alive—the start of the crystallographic analysis of living bodies—the viruses. He himself had been the first to study a virus, that of rabies, from the clinical point of view. In our time, Stanley, Bawden and Pirie have succeeded in extracting them pure and have from this demonstrated their crytalline form. The study by X-rays of these crystals, which I myself have made with my students, has shown that living bodies and pathogenic reproducers, possess at the same time a completely regular arrangement of their constitutent atoms, certainly asymmetric, but essentially similar to those already found in yeasts, biological products without a doubt but divested of reproductive powers.

It will be a vast enterprise to unravel the precise form of this arrangement, but the correctness of Pasteur's idea can already be proved. Molecular asymmetry is directly related to the problem of life.

Pasteur's thought was determined by a clearly fixed idea of the essential difference between living and lifeless things. Possibly he had been a little too insistent on this; he was fighting against the ideas of Wöhler and Liebig, who wanted to reunite the two kingdoms, organic and inorganic. It was one of those dialectical quarrels where both sides are right and wrong at the same time.

Pasteur was absolutely correct in saying that the molecular complexity of living things or even of the products of life were of a completely different order from anything reached by non-living matter. In this, time has fully justified his view. He also considered showing that this living matter was so complicated that according to the laws of symmetry, he could demonstrate that a crystal of albumen could not exist—its symmetry would be so complicated that it could not crystallize. We know now that in this he went too far; proteins do crystallize. Nevertheless, the idea was a good one: the organic world is very complicated, and molecular asymmetry is an intimate and essential part of the organic world. But it is not an exclusive property of life; the existence of right- and left-handed quartz shows that it is possible to have an asymmetry that is not organic.* At first Pasteur had

* Recently a striking and most promising application of this has been made by H. M. Powell using combined chemical and X-ray crystallographic methods. He has shown that synthetic tri-o-thymotide, $3(C_{11}H_{12}O_2)$ has a molecule that can exist in enantiomorphic forms which, however, can easily intercharge (racemize). On crystallizing from a solvent it is possible to secure by starting with a single seed a right- or left-handed crystal at will, the molecules of the other kind all changing to that of the crystal. These crystals, however, nearly always contain in their interstices molecules of the solvent. Since the holes in the crystals are themselves asymmetrical, molecules of one configuration (left or right) will be preferentially picked up. Thus Powell has found a method of separating (resolving) optically active compounds (racemates) without the use—direct or indirect— of asymmetrical substances produced by life. This discovery might have distressed Pasteur at first, but with his passion for facts, however

thought that there must be some universal cosmic asymmetric force which would determine whether a molecule would have a configuration to the right or to the left. Almost imperceptibly he modified this idea in the course of his biological work, replacing it with that of life as a unique chain of beings, each one being asymmetric and each one passing its asymmetry to the next.

Pasteur always insisted on the unique character of the chemistry of life, but Liebig also was right in insisting on the subjection of vital phenomena to the ordinary laws of chemistry. Fermentation depended, as Pasteur showed, on the existence of yeasts, but these chemical reactions themselves depended only on the presence of the dead molecules of enzymes (see pp. 91 f.).

We are now beginning to see that this dialectical opposition of the nineteenth century, which has come up over and over again in the researches spoken of during the course of this anniversary celebration, is explainable in terms of modern structural chemistry. But the central problem is still with us. Even if each molecule of an enzyme or a virus has a determined atomic structure, what needs to be explained is why this particular structure exists in the world out of the billions of possible combinations of the same atoms. This is a question which a mechanical philosophy cannot answer, nor really even formulate. But there is no reason to take refuge in the 'nescio' of mysticism; it is a materialist dialectic which

theoretically unpalatable, he would in the end have been delighted at the human mastery of the shapes of the molecules. This latest example also goes to show that the vein of discovery Pasteur broke open is still far from exhausted.

shows the way forward. The fact of the constitution of proteins is an historic fact. Each molecule in a living entity is connected with a chain of molecules, its molecular ancestors—more or less similar molecules—but as one passes to more and more distant times, they are necessarily simpler and simpler, until they pass insensibly into compounds generated by the light of the sun on the surface of the primitive world. Actual organic substances may thus be called living fossils, and it is in the study of these structures that we will discover, in time, the means of controlling the living world and of understanding its origin.

RESULTS OF PASTEUR'S DISCOVERY

	CRYSTALLOGRAPHY	CRYSTAL PHYSICS	CRYSTAL CHEMISTRY	STEREOCHEMISTRY	CHEMICAL AND PHYSICAL THEORY	
1950		BIJVOET—Absolute determination of configuration of tartaric acid				
1940		UBBELOHDE—Ferro-electric transformations of Seignette salt	BEEVERS—Structure of Seignette salt		BORN, BOYS— Theory of optical activity	
1930		KURCHATOV—Theory of ferroelectricity				QUANTUM PHYSICS
1920		VALASEK—Ferro-electricity of Seignette salt	ROBERTSON, LONSDALE— Structure of organic crystals		HEITLER—Quantum chemistry	
		LANGEVIN—Piezo-electric vibrations	W. L. and W. H. BRAGG— Crystal structure analysis		MOSELEY—Atomic numbers	
1910		VON LAUE—Diffraction of X-rays by crystals			BOHR—Quantum theory of the atom	
			POPE, BARLOW— Chemical theory of crystal structure		RUTHERFORD—Nuclear atom PLANCK—Quantum theory	
1900	SCHOENFLIES, FEDEROV, BARLOW—230 space groups			WERNER—Coordination	THOMSON—The electron RONTGEN—X-rays	ORGANIC CHEMISTRY
1890		CURIE—Piezo-electricity				
1880				VAN'T HOFF, LE BEL— Asymmetric carbon atom		
1870	SONCKE—65 rotation groups			WISLICENIUS—Lactic acids		BACTERIOLOGY
1860						

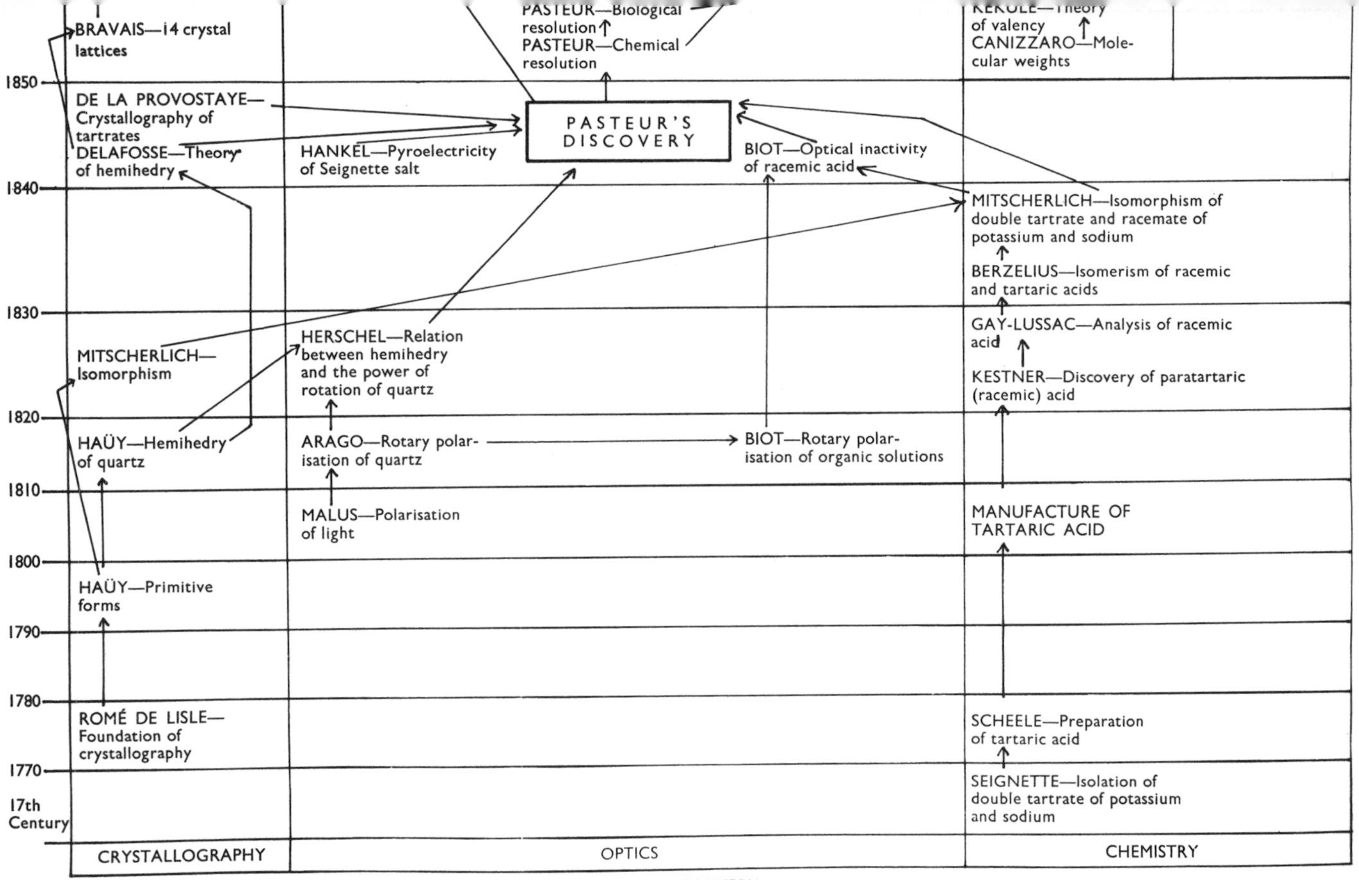

ORIGINS OF PASTEUR'S DISCOVERY

The chart reads (rotated) with a vertical time axis on the left marked:

- 1850
- 1840
- 1830
- 1820
- 1810
- 1800
- 1790
- 1780
- 1770
- 17th Century

Columns at the bottom: CRYSTALLOGRAPHY — OPTICS — CHEMISTRY

CRYSTALLOGRAPHY

- BRAVAIS—14 crystal lattices
- DE LA PROVOSTAYE—Crystallography of tartrates
- DELAFOSSE—Theory of hemihedry
- MITSCHERLICH—Isomorphism
- HAÜY—Hemihedry of quartz
- HAÜY—Primitive forms
- ROMÉ DE LISLE—Foundation of crystallography

OPTICS

- HANKEL—Pyroelectricity of Seignette salt
- BIOT—Optical inactivity of racemic acid
- HERSCHEL—Relation between hemihedry and the power of rotation of quartz
- BIOT—Rotary polarisation of organic solutions
- ARAGO—Rotary polarisation of quartz
- MALUS—Polarisation of light

PASTEUR'S DISCOVERY

- PASTEUR—Biological resolution
- PASTEUR—Chemical resolution

CHEMISTRY

- KEKULÉ—Theory of valency
- CANIZZARO—Molecular weights
- MITSCHERLICH—Isomorphism of double tartrate and racemate of potassium and sodium
- BERZELIUS—Isomerism of racemic and tartaric acids
- GAY-LUSSAC—Analysis of racemic acid
- KESTNER—Discovery of paratartaric (racemic) acid
- MANUFACTURE OF TARTARIC ACID
- SCHEELE—Preparation of tartaric acid
- SEIGNETTE—Isolation of double tartrate of potassium and sodium

INDEX

(The heavy type indicates the main reference to the Subject).

220

Index

Index

Index

Index

Index

Index

Weld, C. R., 139n.
Werner, A. G., 17
Westinghouse Company, 129
Weston, E., 121
Wheatstone, C., 116
Whewell, W., 17n., 164
Whitney, E., 28, 146
Whitworth, J., 27
Wiedemann, G., 64n.
Wilberforce, E. R., 141
Wilde, H., 122, 129
Wilkinson, J., 157
William IV, 143n.
Wine industry, 182, 186 ff.

Wislicenius, J. A., 210
Wöhler, F., 76, 80, 216
Wollaston, W. H., 142n.
Wood, 21, 24
Woolf, A., 46
Working class, 171, 177
Wul, 214
Wyart, J., 195n.

X-rays, 7
X-ray analysis of crystals, 212–16

Yeasts, 218
Young, T., 142n., 161

A selected list of MIDLAND BOOKS

(continued on next page)